CABI CONCISE

SOCIOHYDROLOGY

Sociohydrology is an emergent, interdisciplinary field that addresses the complex relationships between freshwater systems and societies.

Societies use naturally occurring freshwater resources for a variety of (often incompatible) purposes: for spiritual, aesthetic and leisure pursuits, drinking, cleaning, agricultural irrigation, fishing and fish farming, driving hydro-electric systems, cooling data facilities and nuclear reactors and as sewerage conduits. All these utilisations impact natural systems and species in various ways, raising issues of environmental care, sustainability and rehabilitation.

This series seeks to develop and promote research in this field through the publication of research analysing issues such as environmental care, sustainability, rehabilitation, population changes, transformation of waterscapes, water security, biodiversity, human social behaviour cultural representation and more. We welcome proposals covering any of the above issues and/or expanding these approaches into other areas of concern.

Series Editor
Philip Hayward
phayward21c@gmail.com

Dedication

To Paul Rabé, Satya Maia Patchineelam and the River Cities Network teams and administrators for inspiring and facilitating research into urban waterways.

Ghost Rivers and Daylighting in Urban Landscapes

Philip Hayward and Junichi Hasegawa

CABI

CABI is a trading name of CAB International

CABI
Nosworthy Way
Wallingford
Oxfordshire OX10 8DE
UK

Tel: +44 (0)1491 832111
E-mail: info@cabi.org
Website: www.cabi.org

CABI
200 Portland Street
Boston
MA 02114
USA

T: +1 (617)682-9015
E-mail: cabi-nao@cabi.org

A catalogue record for this book is available from the British Library, London, UK.

Library of Congress Control Number: 2026948710

ISBN-13: 9781836994442 (hardback)
9781836994459 (paperback)
9781836994466 (ePDF)
9781836994473 (ePub)

DOI: 10.1079/9781836994473.0000

Commissioning Editor: Claire Parfitt
Editorial Assistant: Theresa Regueira
Production Editor: Theresa Regueira

Typeset by Exeter Premedia Services Pvt Ltd, Chennai, India
Printed in the USA

Contents

Authors

Junichi Hasegawa is a professor at Keio University in Tokyo. He has written books and articles on British and Japanese urban policy and urban planning during and after the Second World War, including *Replanning the Blitzed City Centre* (Open University Press, 1992), '*The Rise and Fall of Radical Reconstruction in 1940s Britain', Twentieth Century British History Vol. 10* (1999) and (with Nick Tiratsoo) *Urban Reconstruction in Britain and Japan*, 1945–1955 (University of Luton Press, 2002). His recent publications include 'The Infilling and Reclamation of Inland Waterways in Tokyo, 1945–1962', *Shima* 17(2) (2023), 'The plans for Tokyo Bay: the challenge of urban policy, 1950s–1990s', *Urban History* 51 (2024) and 'Redeveloping Tokyo's Meiji Jingu Gaien Area: The Metropolitan Government's City Planning Runs Amok', *Asia-Pacific Journal Japan Focus* 23 (2025).

Philip Hayward is an adjunct professor at the University of British Columbia, coordinator of the Flowzones project in north-eastern New South Wales (Australia) and editor of the Environmental Humanities journal Shima. He has published widely on riverine, island and oceanographic topics in journals such as *Anthropocenes, Lagoonscapes, Island Studies Journal, Journal of Marine and Island Cultures* and *Small States and Territories*. His most recent books include *Aquapelagos: Integrated Terrestrial and Marine Assemblages* (2024, Routledge, co-edited with May Joseph); and Blue-*Green Rehabilitation: Urban Planning, Leisure and Tourism in River Cities* (2025, CABI). He is also editor of the CABI Concise Sociohydrology series and is currently editing an anthology on the sociohydrology of islands due for publication in 2027.

Acknowledgements

Thanks to Claire Parfitt at CABI for suggesting the Sociohydrology series and encouraging us to produce this volume. Thanks also to Vincent Bicudo de Castro, Christian Fleury, Ian Stevenson, Satya Patchineelam, Paul Talling and Peter Watts for various research assistances and to Ian S. Jenkins and Francesco Visentin for their valuable feedback on an earlier version of the text.

Glossary

ankyo — a Japanese term for buried rivers and waterways

ankyou — a Japanese term for structural remnants of bridges indicating that rivers once flowed beneath them

autoethnography — a research method that uses researchers' experience of and reflection on encountered phenomena as key elements of their projects

blue-green infrastructure — water, waterside and planted features that mitigate urban heat patterns and ensure non-disruptive runoff of surface waters

blue-green rehabilitation — the rehabilitation of water bodies and adjoining areas of vegetated land to improve water quality, flora and human experience of the areas concerned

buried river — a former surface river flow that has been built over, effectively burying it beneath the surface

culture bombing — a technique of disrupting conventional modes of perception and behaviour

cultural daylighting — the exploration and exposure of the histories and representations of buried rivers

culvert — an artificial passage, such as a tunnel, that conveys water

daylighting — the process by which material built over rivers is removed

deculverting — see daylighting

détournements — cultural performances or texts that modify existing works or structures in an arresting manner so as to open up alternative readings or critiques

ecosocial — connected ecological and social systems and perceptions *ghost river* — an allusive characterization of memories and/or traces of former aboveground flows retained in landscape features, place names, public signage, memories, visual or written records

heat island — an urban area that is significantly hotter than surrounding areas due to its structural materials absorbing and retaining heat

hydrocitizenship — a concept of individuals exercising active stewardship over bodies of water

hydrosocial territories — entities comprising material, social and conceptual aspects of water bodies and human interactions with them

lost river — see buried river

postmemory — 'the relationship that the "generation after" bears to the personal, collective, and cultural trauma of those who came before to experiences they "remember" only by means of the stories, images, and behaviours among which they grew up' (Hirsch, 2025)

psychogeography — an exploration of (usually) urban environments that emphasizes human perceptions and interpretations of spaces, their histories and significance

riparian zone — the area bordering a watercourse or body of water that is directly affected by river flows and/or deposition, also a distinct habitat premised on those aspects

riverine personhood — a concept that regards rivers as having legal and/or cultural personhood and rights

sociohydrology — the interdisciplinary study of the relationship between human societies and water systems of various kinds

subtractive process — developing a building or aspect of the built environment by removing components

urbex — urban explorer(s) or exploration, usually those involved in accessing restricted and/or inaccessible areas (such as sewers) or official and/or decommissioned facilities

Introduction: Interstitial Waterways

Riversides have long been attractive sites for human settlement.[1] They provide fresh water, transport corridors, riparian habitats and fertile arable land arising from riverine deposition. The attractiveness of these locations has also contributed to their degradation, particularly with regard to high-density settlement that has led to the use of rivers for effluent and rubbish disposal. Indeed, Castonguay and Evenden (2012, p. 3) have contended that as the populations of major urban centres such as London, Paris or New York burgeoned in the modern era, developers and city authorities tended to regard rivers as 'mere appendages to waterworks and sewerage systems.' In the 19th and early-mid 20th centuries, in particular, major sewerage programs were undertaken in multiple urban locations, many of which involved the covering-over of sewerage canals and/or functional reassignment of rivers as culverted conduits within wider sewerage networks. Other factors also led to covering-over, including a desire to manage rivers that overflowed their banks and flooded adjacent areas and/or the creation of new surface areas for transport routes or built structures in crowded urban centres.

Adjectives, such as 'ghost,' 'buried,' 'concealed,' 'hidden' and 'lost,' have been applied to the noun 'rivers' in order to refer to stretches (or the totality) of rivers that ran above ground before human intervention transformed them into subterranean flows. The commonly used term 'ghost rivers' is an allusive characterisation of memories and/or traces of former aboveground flows retained in landscape features, placenames, public signage, memories, visual or written records etc. Given the absence of any single compelling term that embraces all these senses, we use the term 'buried' to refer to rivers obscured by physical processes and 'ghost rivers' to refer to subjective and/or socio-cultural senses of buried rivers' lingering presence in the above-ground environment and their current absence from it.

The burials of rivers typify the broader process of the Anthropocene. While almost all aspects of the Anthropocene have been hotly debated, it can be broadly defined as a period when human activities such as extractivism, construction, fossil fuel use and resultant increases in environmental pollution – including the CO_2 emissions that have been a major contributor to global warming – have had a significant impact on the planet. As the *Sustainability Directory* (2026) identified:

> human-built environments – cities, roads, energy networks, agricultural systems, communication lines, and water management structures – now function as dominant geological forces. Their collective mass, reach, and operation fundamentally alter planetary processes.

In this context it is important to note that river burials represent more than superficial modifications of the planetary surface. They remove the micro-climates of riparian zones running alongside rivers and rivers' mitigating effects on urban 'heat islands.' The aquatic biota of rivers is also severely disrupted by culverting and the humans who may once have frequented riversides have been physically alienated from them. In structural and hydrological terms, such burials are also significant for limiting the potential of rivers to channel away flood waters, as impermeable passageways and pipes have fixed capacities and water has to find ways to break out of such constraints during the peak inflow events that are becoming increasingly common as global warming advances.

While the behaviour and management of water have been studied by humans for thousands of years, formal hydrology commenced in the 16th century and became an organized body of study in the early 20th century (Rosbjerg and Rodda, 2019). Despite this, there has been considerable uncertainty over how inclusive the field should be. While Brutsaert (2023) favours a definition that includes terrestrial, oceanic and atmospheric elements, we use the term hydrology in the contemporary sense given by the *Oxford English Dictionary*:

> the branch of science concerned with the water on and under the earth's surface and in the atmosphere, including water cycles and the management of water resources; (also) the features of a region, river system, etc., in relation to this.

In the late 20th century, an increasing recognition of the need to factor social issues, institutions and discourses into hydrological study led to calls from researchers such as Falkenmark (1977, p. 3) to identify humans' relationship with water (and vice versa) as a "complex system of mutual interaction." This area of concern has been characterized as 'sociohydrology' (or, hyphenated, as 'socio-hydrology') and has been developed and promoted by authors such as Sivapalan, Savenije and Blöschl (2012) and Ross and Chang (2020). Herrara-Franco, Montalván-Burbano, Carrión-Mer and Bravo-Montero (2021) provided a valuable service to both hydrology and sociohydrology by conducting a bibliometric analysis of initial publications on sociohydrology in order to evaluate the concept's 'intellectual structure... its conceptual evolution, and its tendencies' leading to them to characterize it as 'a multidisciplinary field, which includes the dynamic reactions and interactions between water and people' based on

> discovery through observation, understanding, and forecasting socio-hydrological phenomena... In addition, it aims to understand and interpret patterns and

> phenomena that arise from rewards in coupled water–society systems as a result of decisions and acts of water management (2021, p. 2).

The authors provide an overview of sociohydrology and its relationship with other science and social science disciplines – as manifested in the literature they surveyed – in Fig. I.1.

Along with the omission of geography from the 'Multidisciplinary Science' category, another significant aspect of the diagram is the lack of any reference to humanities research. While sociology and communication are identified (and might be considered to bridge the distinct fields of social sciences and humanities) anthropology does not appear, along with a range of other relevant disciplines such as environmental humanities, political ecology or Indigenous studies. The most likely explanation for this is that sociohydrology emerged from the natural science context of hydrology and retains many of its paradigms. Researchers from this background are often unfamiliar with other disciplinary paradigms and appear to have been reticent to engage with them. The social science and humanities disciplines referred to above are more commonly associated with a separate approach to essentially similar topics known as hydrosociology, which Wesselink, Kooy and Warner

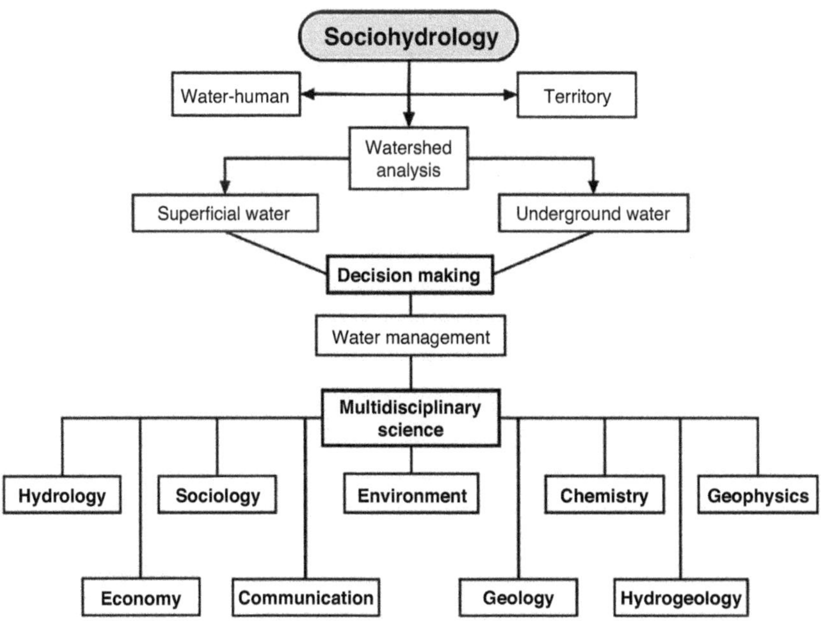

Fig. I.1. Sociohydrology: Conceptual scheme (Herrera-Franco *et al.*, 2021, p. 3 – reproduced with permission of the authors).

(2017, p. 6) identify as having emerged in geography in the early 2010s. While the term has been used by various researchers (see Ross and Chang, 2019), it has achieved less traction and related familiarity than sociohydrology and the concept of the hydrosocial is more often deployed by a range of geographers, humanities and social science researchers than by those who specifically identify *as* hydrosociologists.[2] Boelens, Hoogesteger, Swyngedouw *et al.* (2016, p. 2) have characterized the central concept of hydrosociology as 'hydrosocial territories,'

> the contested imaginary and socio-environmental materialization of a spatially bound multi-scalar network in which humans, water flows, ecological relations, hydraulic infrastructure, financial means, legal-administrative arrangements and cultural institutions and practices are interactively defined, aligned and mobilized through epistemological belief systems, political hierarchies and naturalizing discourses.

In terms of this volume, this schema is clearly relevant for buried portions of rivers, as these are as 'spatially bound, subject-built, socio and natural networks that are produced by actors who collaborate and compete around the definition, composition and ordering of this networked space' (Boelens, Hoogesteger, Swyngedouw *et al.*, 2016, p. 4) as any other hydrosocial territory.[3]

In contrast to the range of material Herrera-Franco *et al* surveyed – and as befits our particular focus – we adopt a hydrosocial approach in this volume that draws on aspects of hydrology, sociohydrology, geography, urban planning, history, art history, anthropology, sociology and literary, media and cultural studies. We thereby attempt the type of dialogue between and synthesis of sociohydrology and hydrosocial analysis that Wesselink, Kooy and Warner (2017) discuss. Our approach also accords with the emergent interdisciplinary fields of environmental and blue humanities, which have developed in parallel (and with various intertwinings) over the last fifteen years. As its name suggests, environmental humanities has sought and facilitated a dialogue between environmental and humanities researchers – as ably represented in key volumes such as Emmett and Nye (2017) and Heise, Christensen and Niemannn (2017). Much of its scholarship has focussed on environmental issues (explicitly or implicitly) considered within the overarching context of the Anthropocene. The journal *Environmental Humanities*, which commenced publication in 2012, is notable for publishing a range of insightful articles on riverine themes ranging from reconciliation and socio-ecological memory in riverine conflict zones (Lyons, 2023) to polluted urban river sediments as 'unintentional infrastructure' (Gonzalez, 2026). While mainly oriented to marine studies, blue humanities – which has been closely associated with its key theorist and advocate, Steve Mentz[4] (2024) – has focussed on the insights on human interactions with the watery spaces of the planet that can be gleaned from cultural representations of them. Mentz's work has also been significant for his recognition of the value of researchers immersing themselves in aquatic environments as forms of 'embodied ecological meditation for the Anthropocene' (2019) that can illuminate their analyses. Oppermann (2023, section 1.1) has also recently identified the need for blue humanities to widen its scope to include

> the study of freshwater ecosystems and biomes, such as lakes, rivers, streams, glaciers, and other freshwater systems. With this expanded horizon, the blue humanities invites a relational aqueous ontology in which salt waters and freshwaters are seen as integral parts of all planetary water ecologies with their co-constitutive and dynamic relationalities.

In this volume – echoing some aspects of environmental and blue humanities – we also expand sociohydrology's and hydrosociology's usual parameters by adding autoethnographical elements to our study of 'buried' and 'resurrected' rivers. Autoethnography is a research method that uses researchers' experience of and reflection on encountered phenomena as key elements of their projects. The method is well established in disciplines such as anthropology and cultural studies and has been the subject of numerous in-depth studies, such as Ellis (2003), Poulos (2021) and the dedicated *Journal of Autoethnography* (2020-). While they were not identified as sociohydrological or hydrosocial projects, two previous academic studies of concealed rivers (Toso *et al.*, 2020; Fisiak *et al.*, 2024) incorporated aspects of autoethnography, with their authors attempting to follow the routes of former surface flows.[5] Lythberg and Hikuroa (2020) take a related but distinctly different approach to discussing the buried Wai-Horotiu stream in Auckland, drawing on their separate ancestries, heritages and life experiences as Pākehā (settler) and Māori (Indigenous) researchers, respectively. Their dialogue explores the extent to which traditional Māori concepts of rivers can inform and be accommodated within mainstream Pākehā thinking and planning (2020, p. 373) and is discussed in the conclusion to Chapter 5. This volume draws on these precedents and a number of related popular urban guidebooks (such as Talling, 2011) to understand the 'afterlives' of closed-over rivers.

Representation, memory and selective rearticulation are fundamental to engagements with buried rivers and the invocation of their cultural ghostliness. The nature and interaction of these components is complex. Indeed, their interrelation creates such a dense web of associations that it is no surprise that memory studies has made limited inroads into sociohydrology to date. Memory studies acknowledges that all memories, including those perceived to be accurate recollections of individuals' direct experiences, are impressions created from various sources and modified in various ways, including their translation into written, visual, audio-visual or other media. If anything, the situation is even more complex with regard to what has been variously referred to as 'social' or 'collective' memory (Sutton, 2020), where the generation of shared 'memories' relies on selectively compiled and inflected aggregates of individual memories, pre-existing social/collective memories and various types of representations of these. The concept of 'postmemory' is also worth alluding to here. Developed by Holocaust researcher Marianne Hirsch in the 1990s[6] it describes 'the relationship that the "generation after" bears to the personal, collective, and cultural trauma of those who came before to experiences they "remember" only by means of the stories, images, and behaviours among which they grew up' (Hirsch, 2025). In this context, she describes its connection to the past as 'mediated not by recall but by imaginative investment, projection, and creation'. Shifting this characterisation to very different type of trauma and erasure, it is possible to identify a type of postmemory as informing many of the projects outlined in the volume.

The concept and practice of daylighting involves the removal of layers of materials built over rivers and, usually, the tidying, re-engineering and/or 're-naturalisation' of their riverbanks. In this regard, the process might be better termed 'open-airing,' given that it involves the decisive opening-up of sections of concealed rivers, rather than a more limited dawn-to-dusk exposure. Despite this, the former term is now well established. Surveying 118 academic publications that addressed the phenomenon, for instance, Khirfan, Peck and Mohtat (2020, p. 4) identified that:

> 82 (or 69 %) explicitly used the root word 'daylight' while the remaining 36 sources (or 31 %) do not explicitly use the term 'daylight' and instead use terms synonymous with stream daylighting. These terms include one or a combination of the following root words: 'deculvert'; 'naturalize'; 'open'; 'reclaim'; 'recover'; 'establish'; 'expose'; 'regenerate'; 'rehabilitate'; 'renovate'; 'restore'; 'resurface'; 'revitalize'; 'bury'; 'uncover'; and 'unearth'. Of the 82 sources, 15 (all published after 2016) exclusively use the term 'daylight' while refraining from using any other synonymous terms while the other 67 sources use, along with 'daylight', one or a combination of the aforementioned synonyms.

Reflecting its prominence in English Language discourse, the term 'daylighting' is used throughout this volume.

Although the daylighting projects discussed in Chapters 4 and 5 might appear to flow smoothly on from the discussions of ghost rivers in Chapters 1–3, it is important to emphasize that the connection is less intrinsic than might be assumed. To begin with, there is no necessary correlation between the intensity of interest in particular stretches of subterraneanized rivers and the likelihood of their flows being deculverted. Indeed, given that the most sustained interest in buried rivers has occurred in densely constructed inner-city environments, these rivers are less likely to be daylighted than those that have had more easily disposable structures constructed over them. There are obvious exceptions – such as the extended section of Seoul's Cheonggyecheon stream discussed in Chapter 4 – but most urban daylighting projects have been more modest in scale and aspiration.

Daylighting can be understood as a 'subtractive process' within the redevelopment of urban environments in the sense developed by Easterling (2003, p. 82), whereby 'subtraction is part of a system of exchange, a function of an active organization of construction and destruction.' He also characterises it as a potentially 'positive tool of space making' that can 'harvest' a 'potent new mixture of ingredients previously separated' in a manner that is capable of 'rewiring an existing ecology or introducing a new instrument in culture' (2013, p. 82). But, in contrast to Easterling's focus on demolition/deconstruction as producing *tablulae rasae*, daylighting has a more functional aim of returning buried rivers and their essential characteristics of flow, cooling and contrast to solid structures to the surface. In that daylighting may be understood as the return of repressed elements to public perception and experience, there is also the question of motivation and effect. If daylighting projects are proposed and deemed worthwhile, which prior moments should be the reference points for them and what are the intended benefits? This is particularly the case with regard to rivers such as the Fleet and Tank Stream that are no longer fed by the water sources that once generated their flows and have been incorporated in urban sewerage systems. Any opening of them to the air prior to their extrication from sewerage networks would principally expose flows of effluent occasionally augmented by rainfall run-off. Even in the unlikely event of official approval for this being obtained, the exposure of such stretches would be positively unattractive for urban populations. Indeed, many successful proposals for daylighting stretches of buried rivers have been asserted within a more generic approach to blue-green rehabilitation of inner-city waterways (Hayward, 2025) rather than reconnection with specific local heritages.

One notable difference between the blue-green rehabilitation of surface flows and river daylighting is that the latter has a more dramatic narrative structure and a more evident climax. The two processes are essentially similar in that they attempt to remediate free-flowing rivers that have become polluted, degraded and/or obstructed. In the case of surface flows, this usually involves the improvement of water quality, the clearance of rubbish and debris from riversides, the restoration of flora and/or the creation of public amenities such as pathways and rest areas. Once initiated, programs usually unfold progressively. While rehabilitation projects can have a narrative climax in terms of an opening ceremony and associated media publicity, this is external to the actual remediation activity. The situation with subterraneanized sections of rivers is different in that work conducted to return them to the surface has much more impact in that the 'lost' is presented as 'found', the 'hidden' is re-exposed, and specific inner-city areas are significantly transformed in the process. The opening up of a buried waterway is thereby a notable (and, in the case of large-scale daylighting projects, momentous) event that defines the subsequent character of the area (at least in the short-term, until the memory of its river once having been buried fades). This dramatic aspect of daylighting substantially explains its appeal to planners, architects and reporters. To assert this is not to undermine the sincerity of efforts to daylight rivers but rather to illustrate that all human actions exist in social contexts in which narrative/event structures are significant aspects of decision making and action.[7]

Reflecting the range of perspectives identified above, our discussions in this volume are multi-facetted. There are a range of rivers buried beneath modern cityscapes, some recalled and celebrated, others largely forgotten.

The first two chapters of this volume provide detailed case studies of two notable buried rivers that now exist as central elements in subterranean water systems. Chapter 1 addresses London's storied River Fleet, an example of an early metropolitan river that has been extensively represented and, more latterly, re-imagined. Chapter 2 examines the related disruption of a river around which Sydney, Australia's first imperial settlement and, later, major city grew, displacing the Indigenous population in the processes. Chapter 3 then analyses the burial of rivers and canals (known as ankyo) in Tokyo, the rise of high profile ankyo explorers and their impact on public discourse and planning. Chapter 4 switches focus to the daylighting of buried rivers and provides an overview of key projects, their motivations and impacts (with a particular focus on Seoul's Cheonggyecheon). Chapter 5 follows this by looking at related concepts of hydro-citizenship and cultural daylighting. The Conclusion reflects on the complex hydrosocial factors and processes involved in remembering and perceiving ghost rivers and rendering sections of them 'alive' again for social experience and/or more general environmental wellbeing through daylighting. Running through the case studies are implicit ontological questions concerning whether and to what extent the original, free-flowing, surface rivers have retained their prior identities in their modern entombed contexts and how they may best be understood in the present. These are key issues for urban planners, communities and cultural researchers as cities are increasingly being reimagined and reconfigured.

Notes

[1] This has been acknowledged by a number of historians and geographers. See Wang *et al.* (2022), for instance, for a detailed study of riverside settlement patterns across China.

[2] It is also worth emphasising that key publications addressing rivers and waterways as cultural heritage 'scapes' (such as Vallerani and Visentin, 2018; Visentin, 2024) do not identify as being either sociohydrological or hydrosociological.

[3] Also see Boelens *et al.* (2025).

[4] Mentz originally referred to the approach as 'blue cultural studies' (2009) but it was subsequently recharacterized as blue humanities.

[5] As Hetherington (2025, p. 668) commented on the work he undertook with Toso and Spooner-Lockey in Montreal, 'we were taken with the strange absent-presence of the Saint-Pierre River. Unsure how to "do ethnography" around a thing that was not really there, we tried to walk its length [and] had incredible trouble doing so' (due to the river's incorporation into a modern urban water system).

[6] See Hirsch (2012) for one clear exposition of the concept.

[7] See, for instance, Cole *et al.* (2022) for perspectives on the role of narrative in decision making, practical action and urban locales.

1 Fleet by Name: The Transformation and Reimagination of an Early Metropolitan River

Philip Hayward

1.1 Introduction

London's River Fleet is one of the most storied ghost rivers on the planet. The various phases of its pollution, modification, burial, representation, socio-cultural rediscovery and re-imagination provide rich examples of the processes involved and of the evolution of a multifaceted cultural entity that remains definitively buried beneath the surfaces of the city. I first became aware of the Fleet in the early 1980s after relocating to London, where I had chanced into a position in the production department of an inner-city advertising agency after graduating from University College, Cardiff. The agency was located in a richly historical area, roughly equidistant between Seven Dials crossroads and Covent Garden piazza. Walking through the built environment of the inner-city, with all its twists, turns, dead ends and various points of access to and perspectives on the Thames, fascinated me. My job involved frequent visits to newspaper premises in nearby Fleet Street, where editing, design and printing processes were then co-located. Conversing with staff and drinking with them in nearby pubs, I heard tales about Fleet Street being named after a heavily polluted river that once ran laterally under it; that after rainfall you could hear it rushing under basements; and that there was a famous play about the river written in Shakespeare's time. These tales were part of 1980s' Fleet Street folklore, or (at least) a form of it related to curious outsiders. At this time, London's ghost rivers had not received the attention that has developed over the last 25 years and I had to comb through various library texts to critically examine what I had been told. Taking the three elements individually, the first characterisation of the River Fleet's route was inaccurate. The original river rose on Hampstead Heath and then flowed south-east, joining the Thames just to the west of Blackfriars Bridge, intersecting with Fleet Street close to St Bride's Church and only running underneath a short section of it. As a result, few – if any – Fleet Street basements could possibly hear the buried river rushing by them and the underground Fleet Street sewer was more likely to have been the source of any noise. As for the play, I was unable to unearth reference to any Elizabethan drama on the topic and have since concluded that the characterisation was a mangled reference to Ben Jonson's mock-epic poem 'On the famous voyage' (discussed further below). But despite my informants' mischaracterisations, their tales intrigued me.

My first (and wholly unsuccessful) attempt to engage with, imagine and represent the Fleet involved trying to document its contemporary point of emergence from the subterranean realm, namely its junction with the Thames under Blackfriars Bridge. In 1981, I had begun to work with Super-8 and 16 mm film, representing various aspects of the city in short films inspired by Walter Ruttman's *Berlin: Die Sinfonie der Großstadt* (1927) and by structural films such as Michael Snow's *Wavelength* (1967).[1] But one production that I never managed to complete (tentatively entitled *Fleet by Name*) provided me with greater insights into its subject than my finished films. This project arose from the glaring disjuncture between a representation of the mouth of the Fleet painted in *c.*1750 that I had encountered and its modern form.

The painting – entitled 'Entrance to the River Fleet as it emerges into the Thames' (Fig. 1.1) – represented the mouth of the Fleet 90 years after it had been canalized as part of the reconstruction of the city after the Great Fire of London in September 1666. Sir Christopher Wren presented King Charles II with a comprehensive plan for metropolitan redevelopment based on what Hebbert (2018, 231) has characterized as 'the best principles of baroque urbanism, with wide straight streets, axial symmetry, monumental endpoints and a waterfront with open quays.' While the King regarded Wren's plan as over-ambitious and prohibitively costly to implement, some elements were approved and constructed,

Fig. 1.1. 'Entrance to the River Fleet as it emerges into the Thames' (oil painting, possibly by Samuel Scott, *c.*1750). Figure used with permission from Bridgeman Images UK.

including the New Canal, a canalisation of the lower Fleet running south from Farringdon to Blackfriars. Wren followed the baroque model in broadening the lower Fleet, constructing wide wharves and installing four high arching bridges whose contours resembled Venice's Rialto Bridge.

The identity of the painting's artist is unknown, but it is commonly attributed to English river- and sea-scape painter Samuel Scott (*c.*1702–1772), or as being in the school or style of that artist. The entrance to the Fleet, visible in the centre of the painting, flanked by full-grown trees on its western edge, is strikingly different from both its modern state and descriptions of its degraded and polluted condition in the 17th and 18th centuries. Aside from the pseudo-Venetian bridge at the Fleet's entrance, there were also other Italianate aspects to the painting that became clearer as I researched the artist. Scott was born and raised in London and sketched its bridges and riverbanks early in his career before falling under the influence of the famous Venetian painter Canaletto (1697–1768). Canaletto is best known for his detailed and romanticized paintings of Venice's canals, buildings and piazzas, which were popular with affluent foreign tourists and collectors. He was a frequent visitor to London between 1745 and 1756 and lived in the city for an extended duration in the early 1750s. During his stays he painted several locally themed works in a similar style to his Venetian paintings.

Scott's oeuvre appears to have been closely modelled on Canaletto's, producing similarly attractive and idealized riverside vistas, and his Fleet painting falls squarely into this category. Bemused by the neo-Venetian scene presented in the painting, I was struck by the Fleet's late 20th-century entry into the Thames via a small passage under Blackfriars Bridge that was largely invisible, except at low tides. I felt a strong sense of poignancy at the manner in which the 'personality' of the Fleet had been effectively effaced in the contemporary cityscape and only emerged, ignominiously, at a largely imperceptible junction. I resolved to try and capture this in late 1982 by filming slow zoom sequences from various positions along the north bank of the Thames intended to illustrate the obscurity of the entry point. But that very obscurity (see Fig. 1.2) – and my inability to get onto adjacent rocky shores on rain-free days during equinox tides, meant that all I generated were visually dull, sideways perspective sequences that would have been inexplicable without an accompanying text. In this manner, the Fleet remained elusive, frustrating my attempts to open it to scrutiny and reappraisal.

1.2 The Decline and Representation of the Polluted Fleet

The elusive nature of the Fleet's entrance to the Thames was all the more frustrating as it seems broadly agreed that the Fleet was named after its wide mouth and the particular anchorage it offered to vessels (*fleót* being the Middle English term for such a locale). Indeed, in early usage, the term applied specifically to the river's mouth with the overall river being referred

Fig. 1.2. The entrance to the Fleet under Blackfriars Bridge (bottom left-centre of image, partially obscured by shadow) (by Mark S. Jobling, 2002, Wikimedia Commons).

to by names such as the River of Wells (on account of the number of these along its upper reaches). The river was once navigable as far upstream as Kentish Town (5.5 km from its junction with the Thames) and was a thriving waterway during the Tudor period (1485–1603). But while the river might have been an appealing feature to those living along its banks in the mid-late Medieval period, as the city's population ballooned, the volume of sewerage, offal and general refuse deposited in it became an increasing cause of olfactory discomfort and concern. Ships navigating its lower stretches found themselves floating in what was effectively a shallow aboveground sewer that was fed by ditches that ran down the middle of roads such as Fleet Street. This aspect was vividly represented by poet and dramatist Ben Jonson in his poem 'On the famous voyage,' which dates from around 1610 and dwells on riverine pollution and the more general malodorous squalor of the area. The urban waterscape represented in the poem was one that Jonson was familiar with. In the early 1600s he lived in the Blackfriars area, on the east side of the Fleet as it approached the Thames. He set his play *The Alchemist* there and his satirical work *Eastward Ho!* (co-authored with George Chapman and Ben Marston) was staged at the Blackfriars Theatre in 1605. The Fleet and its surrounds, as featured in the poem are, therefore, a real place interpreted in a dramatic manner.

The voyage identified in the poem's title is both a linear one, travelling along a stretch of the lower Fleet, and an allegorical one, with the area of

London concerned being represented as a human body marked by its coprological character and function and characterized by its stench and excrescence. This is evident in Jonson's colourful characterizations, which feature repeated reference to the faeces present in it and switches between classical allusion and prosaic observation:

All was to them the same, they were to pass,
And so they did, from Styx, to Acheron:
The ever-boiling flood. Whose banks upon
Your Fleet Lane Furies; and hot cooks do dwell,
That, with still-scalding steams, make the place hell.
The sinks ran grease, and hair of measled hog

Despite the Great Fire of London and the baroque-influenced canalisation of the lower Fleet that followed, problems with the draft of the canal (due to infilling from deposition and dwindling upstream water supply) meant that it soon became heavily polluted again. Its dire state in the early 18th century was captured in Jonathan Swift's 1710 poem 'A description of a city shower,' which detailed that sewerage channels running from Smithfield meat market and from the area of St Sepulchre's Church converged in the Fleet:

And in huge confluent join at Snow Hill ridge,
Fall from the Conduit prone to Holborn Bridge.
Sweepings from butchers' stalls, dung, guts, and blood,
Drowned puppies, stinking sprats, all drenched in mud,
Dead cats and turnip-tops come tumbling down the flood.

The lower stretches of the river – then, in its degenerated form, commonly known as 'The Fleet Ditch' – also feature in Alexander Pope's mock-heroic, narrative verse 'The Dunciad' (published in three versions between 1728 and 1743), a satire on the aesthetic and moral deterioration the poet saw as afflicting London in this period. But the most extreme representation of the river was presented in Arthur Murphy's poem 'An Ode to the Naiads of Fleet-Ditch' (1761). One of the most effective elements of the poem is its insertion of naiads (water nymphs), more commonly associated with freshwater streams, pools and springs, in a river clogged with faeces and offal. The opening verse is graphic in describing the flows of sewerage into the Fleet:

The nut-brown Naiads of that fable flood,
To which auxiliar sewers their homage pay,
And little rills, meanderings o'er the mud,
Wind from a thousand urinals their way.

In all likelihood reflecting the association of Blackfriars and adjacent areas with prostitution in this period, the poet provides a less than fragrant description of the naiads' charms, describing them as 'sitting on the swampy bank/ with fabled ooze, your tresses dank.'

Despite the progressive burial of the Fleet as the city expanded north, open sewerage ditches ran along several main streets close to its lower course until the mid-1800s, when they too were concealed.[2] A particularly detailed representation of the processes involved was published in *The Illustrated London News*, (Fig. 1.3) on 4th October 1845, which summarized that:

> The Sewer of Fleet-street, the subject of our Engraving, having been found insufficient to carry of the water, for which it was intended, it became necessary to cut deeper, and construct a new sewer: the greatest requisite depth is 25 feet, which decreased to about 17 feet near Temple Bar; hence the Sewer runs easterly to Water Lane, where it is joined by another Sewer, which runs into Whitefriars Dock.

While the Fleet and its tributary sewerage channels were removed from public gaze and olfactory perception by the mid-1800s, they still deposited their effluent in the Thames and contributed to the extreme pollution of waterways that

Fig. 1.3. 'Deepening the sewers in Fleet Street'; *The Illustrated London News*, 4 October 1845. Construction work to deepen the sewer in Fleet Street, with the Old Temple Bar (which was demolished in 1878) visible at the rear. (Wikimedia Commons, 2025).

resulted in the so-called 'Great Stink' of the summer of 1858, when royalty, parliamentarians and residents experienced nausea and discomfort from the smell of sewerage fermenting in the Thames during a period of unusually high temperatures (Halliday, 2001). This led to the speedy passage of 'The Metropolis Management Amendment Act' (United Kingdom Parliament, 1858) that empowered the Metropolitan Board of Works to solve the problem. This task was assigned to their chief engineer, Joseph Bazalgette who created the basis of London's modern sewer system and substantially reduced the sewerage content of the section of the Thames running through the city by pumping the majority of the effluent down into the river's lower reaches (with a predictably poor outcome for the water quality and riverside population of these areas).

With the completion of the works that buried sewerage ditches below ground around the Fleet in the 1860s, the subterraneanisation of the river and its tributary sewers was complete and its surface flow slipped into memory, principally recalled by place names – such as, most obviously, Fleet Street – and through a commemorative volume entitled *The Fleet: Its river, prison and marriages*, written by historian John Ashton in 1888. This text, which has provided the basis for many subsequent accounts of the river's history, is essentially an historical miscellany about the river, providing a history of its sources, tributaries and flows, developments along it, famous people associated with it and notable events. Ashton's volume is significant for both the amount of historical detail it includes and for anticipating later investigations of and attempts to connect with the river's previous aboveground form and the lives of those who relied upon it. As its Introduction states:

> Time has wrought such changes in this land of ours, and especially in its vast Metropolis, "The Modern Babylon," that the old land-marks are gradually being effaced – and in a few generations would almost be forgotten, were it not that some one noted them, and left their traces for future perusal. All have some little tale to tell; even this little River Fleet... are things utterly of the past, entirely swept away, and impossible to resuscitate, except by such a record as this book. (Ashton, 1888, p. vii)

As the following section outlines, while the Fleet has not been physically resuscitated in any way, remaining entirely 'un-daylighted,' it has been increasingly researched and traces of its surface course have been mapped, walked and described in manners that underline the value of works such as Ashton's for hydrologists, social historians, heritage researchers, town planners, residents and tourists alike.

1.3 Re-engaging with the Fleet

One of the most frustrating aspects of writing this chapter has been the lack of information about attempts to research, memorialise and/or trace the surface routes of London's buried rivers in the early mid-20th century, aside from Alfred Foord's volume *Springs, Streams and Spas of London* (1910), which the author identifies as collating data from 'innumerable books, pamphlets and magazines, as well as the columns of newspapers' from the 16th to early 20th centuries (Foord, 1910, p. 15). The latter phrase suggests that there was some continuing interest in the topic in the first decade of the 20th century but offers no clues as to what form that may have taken. The next 50 years is even more of a mystery. It is difficult to ascertain the extent to which interest persisted through the First World War, the Great Depression, the Second World War and the immediate post-war period. But some indication that it *did* persist is provided in the introduction to Nicholas Barton's *The Lost Rivers of London* (1962), in which the author states that he encountered a 'considerable amount of literature, and an even greater amount of interest' in the topic while working on his volume and became aware of a number of people researching similar themes (Barton, 1962, p. 12). Barton's book succeeded in being reviewed in high-profile publications such as *The New Statesman* (Haycraft, 1962) and the *Tribune* (Groves, 1962), with the latter's review opening with the plaudit that:

> This is a book to welcome and praise, for it is a fascinating and painstaking record in prose, picture and map of London's lost and buried waterways, one to be enjoyed as much by the ordinary reader as by the specialist. (Groves, 1962, p. 10)

While Barton's book did not immediately stir a popular movement towards acknowledging and exploring the city's lost rivers, it was received well-enough to be republished in 1982 and to be expanded in a version co-authored with Stephen Myers in 2016.

The momentum that Barton's research created was enhanced and inflected in the 2000s and 2010s by a small group of British writers and urbanists who were inspired by radical French philosopher–provocateur Guy Debord's concept of the *dérive* – a perambulatory 'drift' through urban landscapes undertaken by a small group of individuals in dialogue with each other[3] – and his contention that such activities could produce 'psychogeographies' of place that could reveal the latent content of marginal locations and how various sets of elements are superimposed over others. Drawing on this, Lyons (2017) has characterized that:

> The transition of a space from one use to another undergirds much of psychogeography's preoccupation; the notion of a palimpsest – an object or piece of writing with new material superimposed over earlier writings – is particularly important… This is what makes psychogeography a particularly useful critique beyond mere urban re-enchantment. Psychogeography thrives as an interrogation of space and history; it compels us to abandon – at least temporarily – our ordinary conceptions of the face value of a location, so that we may question its mercurial history.

Her characterisation illustrates the attractiveness of Debord's concepts to those urban 'drifters,' architects and urbanists who seek to experience and explore the palimpsestic nature of urban environments and what (literally and figuratively) lies beneath them. In this simultaneously holistic and interrogative perspective, elements such as subterraneanized rivers can be regarded as just as much part of the multi-layered reality of urban spaces as current structures and just as worthy of consideration.

The author Iain Sinclair was widely associated with psychogeography in the early 2000s, producing a personal, literary variant that embraced traditions such as Swedenborgian mysticism, which perceived correspondences between the spiritual and material realms. He has devoted particular attention to London's subterraneanized rivers and has regularly returned to them, expressing his conviction that 'walking over or alongside the buried rivers of London stitches a form of collective memory into our sides' (2018, p. 3). Given the complexity of place memory outlined by theorists such as Sutton (2020) , we might understand this as more of a process of embroidery – understood in its sense of creative interpretation and elaboration – than acquisition. Discussing what he terms the rivers' 'haunting complexity,' he has declared:

> They're not lost, not at all. Just because you can't see a thing... doesn't mean that it's not there. The rivers continue, hidden and culverted as they might be, to flow through our dreams, fixing the compass of our moods and movements. The Walbrook, the Fleet, the Tyburn, the Westbourne, the Effra, the Neckinger: visible or invisible, they haunt us... (Sinclair, 2018, p. 5)

Critics such as Bonnet (2009) have interpreted this aspect of Sinclair's work as mapping 'modernity and loss as an engagement with alienating, often brutally instrumental landscapes' (2009, p. 54). Sinclair's work – and particularly *Swimming to Heaven: The Lost Rivers of London* (2013) – is marked by the deftness with which he handles these elements. London's 'not lost' subterraneanized rivers are enticingly polyvalent entities in this scheme, evoking the complexity of what has been hidden from the modern city, what runs beneath it and what might be revitalized.

Sinclair's concerns also overlap with those of contemporary urban explorers concerned to descend into subterraneanized flows. While they can be used more generally, the terms 'urban explorer' and 'urban exploration' have a specific contemporary meaning with regard to unauthorized access to buildings, industrial sites and underground facilities of various kinds. This activity – sometimes abbreviated as 'Urbex' – developed as a conscious project in the 1980s and appeals to its afficionados by combining the transgressive activity of trespass with access to and experience of restricted environments. While by no means all its adherents follow conscious agendas, the Urbex website (2025) has identified that:

> urban explorers often cite psychogeography as an influence, framing their adventures as not just adrenaline-fuelled trespassing but also a critique of modern life – reclaiming parts of the city that are normally ignored or off-limits. In a sense, exploring a disused railway tunnel or an abandoned office block is a way of saying *"the city belongs to us too, not just to authorities or developers."*

Journalist Peter Watts documented a visit to the submerged Fleet facilitated by two urban explorers (referred to as Stoop and Jondoe) in 2011:

> It's cramped, damp, dank and dirty but doesn't smell too bad. Stoop and Jondoe glide like skaters along the slippery floor while I splash clumsily behind, using slimy walls to keep my balance. We head downwards along a series of mini-waterfalls. The light and noise that intermittently emerges from the grates overhead suggest we aren't getting deeper, but simply following the gradient of the road... The tunnel is around ten feet tall and wide, so we can walk two abreast. It's about the same size as a tube tunnel. The smell slowly subsides, although lumps of faeces and toilet paper gather in places where they've washed against the brickwork. Otherwise, there's just a trickle of brown water ferrying the odd cotton-bud downriver... After about two miles of trudging, we emerge into an enormous end chamber, more than 20-feet high and elaborate in design and construction. Two short tunnels lead from here, ending in huge metal flaps, which we assume open directly into the Thames.

Since Watts' brief encounter with the lower stretch of the Fleet in 2005 there have been a number of other informative representations. In 2016, for instance, writer and academic Tom Bolton and a video crew working with the *Londonist* website participated in an officially approved visit into a section of the Fleet running under Camden and produced the short video *Inside the River Fleet Sewer*.[4] More recently, solo urban explorer Valdigger has shot unauthorized video sequences that show the Fleet's lower subterranean passages (such as *River Fleet Connection to the Super Sewer*, 2025[5]). These productions allowed the hidden to be perceived online after over 150 years of restricted visibility and a related inability to imagine its flow and brick-lined confines.

Writer, filmmaker and psychogeographer John Rogers, who co-directed the documentary *London Overground* (2018) with Sinclair, has been another significant contributor to the investigation of the paths of the Fleet and its traces in the modern cityscape, particularly through his two video documentaries *Walking the River Fleet* (2020)[6] and *Walking forgotten tributaries of the River Fleet* (2025).[7] The former opens with a characterisation that the Fleet is 'London's most famous, most notorious and most mysterious lost river' and represents a 1-day walk along its route in a 40-minute-long video commencing with a trickle of water at the Vale of Health in Hampstead and ending at the Thames. The latter offers informative footage of streets following the routes of submerged tributaries of the Fleet and also recounts Rogers' meetings with various walkers independently tracing the same routes, illustrating the number of those involved in the practice. While G. H. Davies's 2013 e-book, *Following the Fleet: London's Secret River*, usefully updated various aspects of Ashton's 1888 volume, some of the most creative engagements with the Fleet's past have arisen from Rebecca Leach Mcdonald's attempts to animate its lost history through river trace walks. These have included attempts to revive the river's former reputation and designation as the 'River of Wells' by inviting walkers to wish on jars of water filled from Goodison Fountain on Hampstead Heath.

The psychogeographically informed work detailed above has been accompanied by a body of less theoretically inflected material on London's network of buried rivers. Paul Talling's 2011 volume *London's Lost Rivers* and the author's subsequent organization of related walking tours (including over 200 of the lower Fleet area, to date) have been particularly significant in the 21st-century revival of interest in the topic. Like the authors of the previously discussed volumes, Talling has no formal training in hydrology but the thorough nature of his research and the extensive photo-documentation featured in his volume secured it favourable reviews and radio and TV coverage.[8] Author and academic Tom Bolton has also produced two similar publications intended to provide detailed routes for urban explorers to follow (2011 and 2019) and David Fathers recently published a ghost river–themed guide (2024). Continuing interest in the topic was also evident in the staging of the 'Secret Rivers' exhibition by the Museum of London Docklands in Summer 2019. The creation of a Facebook page dedicated to 'London's lost rivers, undergrounds sewers and stinkpipes' by and for enthusiasts in 2019[9] underlined community interest, which was also evident by the success of organized 'secret river' walks run by the Inner London Ramblers Association in 2020–2022, and the proposal of a network of new walking tracks along their routes (which has not come to fruition).

While the extensive construction over the majority of the Fleet's former surface route appears to have deterred aspirations to daylight sections of it, one proposal submitted by London's 4orm architectural firm to the Royal Institute of British Architects' 2013 competition 'Forgotten Spaces' merits discussion. The competition invited architects to propose refreshments of neglected, inner-city locales. Inspired by Ian Sinclair's observations about St Pancras Old Church as a 'marooned and discarded' site on a mound 'above the now submerged and piped Fleet River' during extensive construction work on the Kings Cross Central redevelopment[10] that encircled it (Sinclair, 2005, p. 16), 4orm's entry proposed the recreation of a section of the river running under a public garden adjacent to the church. Characterising the park as 'one of the places along the course of the Fleet where the topography of the river bank can still clearly be recognized' (Gooden, 2019, p. 138), the firm contended that the location provided 'the opportunity to create a place of delight, recreation and relaxation and to give a fitting setting to the historic church and gardens' (Gooden, 2019, p. 138). Rather than an actual daylighting of the buried Fleet (whose current water quality and reliability of flow are uncertain), 4orm's proposal harked back to approaches to hydro-engineering environments pursued by 18th-century English landscape architects such as Capability Brown.

> In the future, it may even be possible to once again supply this resurrected watercourse from the clean waters of the Hampstead ponds. For this proposal, however, as happens at the Serpentine in Hyde Park, water will be drawn from boreholes deep into the chalk aquifer, with the pump houses concealed in sham bridges[11] through which the river flows at its northern and southern ends. The use of the sham bridge recalls the English landscape garden tradition, where a watercourse is extended visually, and in the imagination, to a suggested world beyond. (Gooden, 2019, p. 139)

Note the open embrace of artifice here. The focus is not on any notional recreation of original pre-metropolitan, blue–green spaces but on an ingenious means of evoking them.

One of the most striking aspects of 4orm's entry was the use of a collage that suggested the riverscape they envisaged (Fig. 1.4). This featured figures and details from George Seurat's 1884 painting *Bathers at Asnières* inserted into a visualisation of the 'resurrected watercourse' proposed to run through the public garden. Richard Gooden, one of the architects involved in the proposal, explained that they took the well-known work

> and adapted it to describe the changes at St Pancras. Seurat's painting juxtaposes a bucolic foreground scene with a distant context of the industrial world – factory chimneys and a steam train crossing the bridge – that the modern reader knows will soon advance on and overpower its location. In contrast, our proposal suggests a reversal of the ineluctable overthrow of the natural world by the industrial city. To reinforce the idea of reversal, we chose to mirror the figures from the Bathers so that they now face in the opposite direction – as does the train on the viaduct (a Eurostar service rather than the steam locomotive of the 1880s). It is intended to suggest that the river can reclaim the city. (Gooden, 2019, p. 139)

Gooden's discussion of the collage gives a strong sense of purpose to the firm's entry, proposing a 'reversal' of industrial urbanism complementary to aspects of the nearby Kings Cross Central redevelopment (such as the fountains in Granary Square and tiered pools in Pancras Square[12] – Fig. 1.5) and ambitiously identifies the proposal as a small step to humans 'reclaiming' the city.

While 4orm's proposal never proceeded past its shortlisting in the competition,[13] it chimes with similar optimistic visions for the future of the area. Chris

Fig. 1.4. Collage of Seurat's *Bathers at Asnières* and a visualization of the redesigned public garden. (Figure used with permission from Richard Gooden, Stephen Coleman and Claire Priest of 4orm Architects.)

Fig. 1.5. Water feature, Pancras Square (design by The Fountain Workshop, photograph by Peter Wood, Wikimedia Commons, 2018).

Douglas, Senior Policy and Projects Officer at London Borough of Camden, for example, recently concluded a report on a proposed walkway along the river's former surface route through the borough with the assertion that such a 'blue corridor' along a daylighted Fleet would provide a major asset.

> I like to imagine that one day the Fleet might flow through Camden again: sparkling streams with riverside bars and cafes, fish swimming down the river and rich green flora around. For now, I think of Pancras Square's cascading ponds outside our offices as a secret nod to the Fleet flowing just yards away, hinting at what once was and might be again. (Douglas, 2025)

Douglas's imagination of a restored and revitalized Fleet flowing through Camden, populated by fish and surrounded by verdant riversides, is an idyllic one. While it is not inconceivable that such a project could be planned and/or implemented, its costs would, in all likelihood, be prohibitive in such a densely developed and high-value real estate area as Camden. Indeed, Douglas's vision invites comparison to Alistair Bonnett's characterisation of British psychogeography as a 'creative space in which feelings of loss and redemption are explored and negotiated' (2009, p. 54). At time of writing – i.e. early 2026 – the river's fate (like its surface) appears firmly sealed, but given both its varied history and the increasing risk of intense precipitation events during a period of global climate change, may be subject to further transformations. Indeed, history suggests that such events may be more than likely. Prior to the Fleet's relegation to a subterranean flow, its riversides had been subject to major flooding events in the 13th, 14th and 17th centuries (UCL River Fleet Restoration Team, 2009, p. 7). In one event in 1846 that has been largely omitted from many modern literary reinscriptions of its history, an intense storm caused such an inrush of water into a section that it was reported to have *exploded* (Thornbury, 1878), rapidly flooding cellars and residential streets. Given that many of the current channels within the Fleet system are at least 150 years old, it is questionable whether they can safely contain the volume of water that might be channelled into them during peak weather events and similar explosive outbreaks may eventuate. If occurring at sufficient scales and intensities, these may even cause impromptu *subtractions* of existing surface structures (Easterling, 2003) that might necessitate newly engineered environments around the river and its tributary sewerage channels and fresh imaginations of the river's presence and potential in a 21st-century Anthropocene environment.

Notes

[1] These were screened at open evenings and summer shows at the London Filmmakers Co-Op in Camden in 1981 and 1982 and are not currently in distribution.
[2] Also see van Lieshout (2024) for a discussion of iterative processes that preceded major water engineering projects in the capital.
[3] See particularly Debord (1958).
[4] https://www.youtube.com/watch?v=ioRcT9eElAo
[5] https://www.youtube.com/watch?v=YIDVpdG-YaQ
[6] https://thelostbyway.com/2020/11/walking-the-river-fleet-londons-lost-rivers.html
[7] https://thelostbyway.com/2025/09/walking-forgotten-tributaries-of-the-fleet.html
[8] Archived online at: https://www.londonslostrivers.com
[9] https://www.facebook.com/groups/521714131710980
[10] See Argent St. George (2001) for a statement of the project's guiding principles and Bishop and Williams (2019) for a critical survey of the project.
[11] Sham (i.e. imitation) bridges were decorative features in 18th century landscape design (with a well-known example being at Kenwood House on Hampstead Heath). Goodwin is suggesting concealing pumphouses in such structures.
[12] David Bracey, director of the design company The Fountain Workshop has identified that 'Granary Square was once a canal basin, where boats and barges delivered grain to London for distribution to millers' and that the firm were inspired by this aspect in designing water features for the area (personal communication, March 24, 2026).
[13] See Bould (2013) for an insightful review of the 26 shortlisted projects and his contention that through their speculative nature 'each display conjures a possible future that... is never seriously intended to become reality. (Meanwhile, dancing just out of reach are reputation and the possibility of actual commissions).'

2

Sydney's Tank Stream: The Submergence of an Indigenous Riverscape and Its Cultural Traces in a Metropolitan Centre

Philip Hayward

2.1 Introduction

Between the mid-16th and mid-18th centuries, London's population grew from *c.*80,000 to over 750,000 as it became the centre of the British Empire, the world's largest city and its largest port. But despite the success of Britain's imperial enterprise in this period, the country was also facing a rapidly changing geo-political situation as its North American colonies revolted against British rule and as it sought to consolidate its global power via acquiring new territories. Following Captain James Cook's exploration of Australia's east coast in 1770 and his claiming of it for the United Kingdom, the continent provided one such opportunity and plans were made to take advantage of it. In 1787, a flotilla of ships carried around 1310 (predominantly male) convicts, marines, officers, administrators and voluntary settlers to Australia to establish a penal colony. Arriving in January 1778, the flotilla initially visited Botany Bay before sailing a short distance north, entering an extensive ria (drowned river valley) system and discovering a more promising site that they named Sydney Cove. Despite being inhabited by an Indigenous community whose consent was not sought, the area became the nucleus of the future city of Sydney and for the subsequent British conquest of the continent.

At the time of first contact, the area around the cove was inhabited by the Gadigal, one of 29 clans of the broader Eora nation who inhabited what is known today as the Greater Sydney region. Sydney Cove appealed to Governor Phillip on account of the presence of a short, wide stream deep enough for ships to enter. Exploration of the vicinity established that the stream (simply referred to as 'the streamlet'[1] by the early colonists) originated around a kilometre inland, in a swampy area close to the present-day Hyde Park, then descended about 30 m, running north through a series of low cascades and being supplemented by several small springs, before draining into the cove. Sparse accounts from the first phase of settlement suggest that the area through which the stream flowed was densely wooded (Fig. 2.1) and Macphail and Owen (2018) have characterized that this comprised 'casuarina swamp forest in the estuary ... or stands growing along the lower reaches of the stream, with ground ferns dominating damp sites' (2018, p. 25). The authors also concluded their study with the characterisation that there is 'a strong probability' that the area around the streamlet was part of 'a cultural landscape shaped and managed in particular ways by the Aboriginal population over millennia of skilful burning' (2018, p. 26). It is also increasingly acknowledged that Indigenous Australians have a deep spiritual connection with and perceived duty-of-care for water bodies. This has been summarized by the National Cultural Flows Planning and Resource Committee (2017) in the following terms:

> For Aboriginal people, water is a sacred and elemental source and symbol of life. The resources provided by aquatic ecosystems are a pivotal part of spirituality and the cultural economy. The rivers are the veins of Country, carrying water to sustain all parts of the landscape. The wetlands are the kidneys, filtering the water as it passes through the land.

While the original Gadigal name for the area's stream is not known, there is archaeological evidence that stone tools were made on its banks and that the

Fig. 2.1. 'Sydney Cove, Port Jackson' by William Bradley (1788) (Wikimedia Commons), with the entrance to the streamlet later known as the Tank Stream at centre of the image.

area was also used for burials (Irish and Goward, 2013). In the early phase of settlement, the clan withdrew from the colonized area around the cove and avoided contact with the newcomers who had rapidly set to clearing vegetation and constructing buildings of various kinds. Retaliatory violence was also conducted on settlers around the fringes of the settlement but by the mid-1800s, their numbers reduced by introduced diseases they had no established immunity to, remaining Gadigal clan members had largely retreated south towards Botany Bay.

2.2 Transforming the Streamlet

The use and maintenance of the streamlet was an immediate concern for the colony's first governor, Arthur Phillip. Phillip was born in 1738 in the Bread Street ward of the city of London, half a kilometre to the east of the Fleet River, where he lived until the age of 15, when he crossed the Thames to study in Greenwich before joining the Royal Navy (Fletcher, 2006). As detailed in Chapter 1, the Fleet was heavily polluted at this time and was a cause of considerable discomfort and concern to residents. While there is no evidence to confirm the conjecture, it seems likely that Phillip's attempts to manage the Sydney Cove stream in the early phase of the colony reflected his experience of severely degraded rivers in London and his wish to avoid these becoming established in the new colony. McIllwraith's (1952) detailed outline of the management of water resources during his governorship identifies that while settlers were allowed to occupy areas of cleared land adjacent to the stream, the fresh water source was 'jealously guarded,' with a 'green belt 50 feet [c15 metres] wide' being 'preserved as a reserve' within which 'cutting of timber was forbidden' and 'stock were not permitted to graze' (1952, p. 88). This appears to be one of the first documented accounts of such a strategy in modern urban planning and merits greater acknowledgement in that regard. Despite this, Phillip soon had to contend with another issue with regard to the unreliability of fresh water supply from the stream.

During the dry summer of 1789–1790, the stream's flow dwindled to a trickle and Phillip ordered the construction of three water storage tanks (capable of holding up to 20,00 litres each) in the sandstone bedrock adjacent to the mid-area of the stream.[2] The water for these was channelled from the streamlet and stored for subsequent use, with it being primarily regarded as a supply source for these and increasingly being known as the Tank Stream on that account. Phillip returned to England in 1792, leaving successive governors to deal with increasing pollution pressures on the stream and, thereby, the tanks it fed, as the population of the colony increased. In 1803, Governor King issued an order prohibiting the throwing of 'any filth into the stream of freshwater, cleaning fish, washing, erecting pigsties near it or taking water out of the tanks' and also ordered the construction of fences along the stream (McIlwraith, 1952, p. 41). Despite such efforts, substantial clearance of the stream's banks was undertaken and the stream continued to be polluted with sewerage and farm and abattoir by-products. Indeed, by 1836 the water quality had become so poor that Governor Brisbane issued an order forbidding anyone to drink from it.

One of the clearest impressions we have of the stream in this period is provided by English artist John Skinner Prout's watercolour painting 'The Tank Stream, Sydney,' thought to have been painted in 1842. His watercolour is insightful. While retaining elements of a picturesque aesthetic (particularly in its mid-rear planes and sides) that might have appealed to a local market,[3] his painting vividly chronicles the stark streamside resulting from clearance of its vegetation. Governor Phillip's attempts to retain a green belt around the stream had clearly been in vain and the painting shows denuded banks and piles of cut timber in the lower left foreground and middle right midground. A small patch of low, green vegetation features at the rear of the latter, and while there are individual trees on the left and right sides of the vista, the overall impression, heightened by the muted brown–beige colours, is of a mostly bare landscape. With the stream running through its centre, the painting also neatly shows the division between the working-class area of the west bank and the leafier and pastured east bank. Overall, there is a drear sense of purposelessness to the scenario, painted in a period when New South Wales (NSW) was experiencing economic depression and a sense of the stream being a neglected tributary to the majestic harbour that lies beyond at the rear of the painting, where a schooner floats just offshore.

While alternative fresh water sources were found in the form of wells and a constructed channel bringing water from ponds to the south-east of the city, continuing concern about exposed sewerage flowing in the Tank Stream and related outbreaks of disease led the city authorities to begin the drainage of the (seasonally) swampy area that fed the Tank Stream and to investigate the construction of a sewerage system and the provision of suitable pumping systems for it. The project was led by a newly appointed head engineer, Edward Bell, a Londoner who had relocated to Sydney from England in 1856 at the age of 44. One (presumably unforeseen) consequence of the drainage of the stream's source was almost total reduction of freshwater flow through the Tank Stream, leaving sewerage and irregular rainfall as its main fluid inputs. In order to conceal the corrupted flow, part of the stream was diverted under Pitt Street in 1856–57 (Fig. 2.2). In 1860, a further 207-metre-long section was covered over with an arched sandstone roof, and two remaining sections were covered over with brick arches in 1870 and 1880. With the extensive sewerage works that commenced in the city in 1857, sewerage flowing down the subterraneanized Tank Stream was diverted away from the Harbour, only overspilling into it following high precipitation. (McIlwraith, 1952, p. 50)

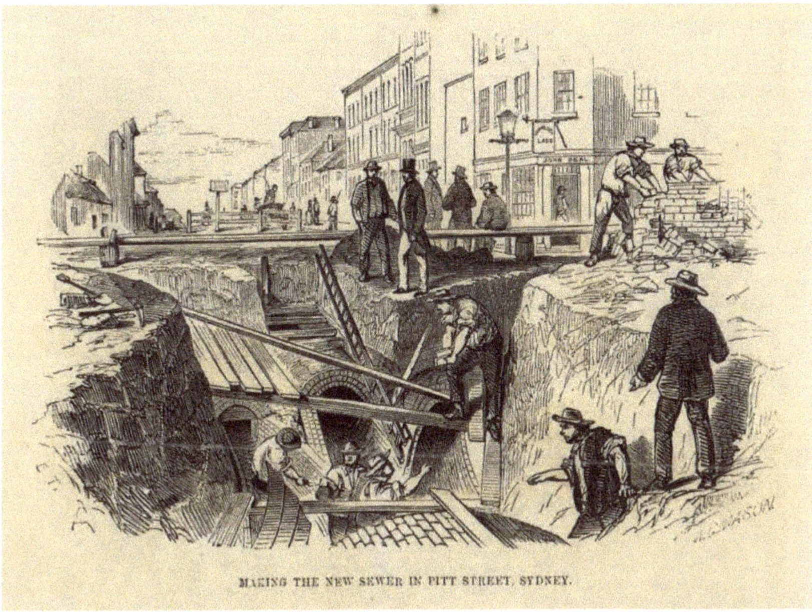

Fig. 2.2. Walter J. Mason's 'Making the new sewer in Pitt Street' (1857). National Library of Australia. [nla.pic-an8021132]

2.3 Tracing the Tank Stream

I first became aware of the Tank Stream in 1989, shortly after moving to Sydney to take up a position at Macquarie University. Wandering around the area south of Circular Quay one hot summer evening, I came across a bar named the Tank Stream. I had no idea what the name meant and (to my recollection) there was no signage or information that gave any background. Walking away from the venue, I also noticed that the name of the small street it was located on was Tank Stream Way. Mentioning this visit to a long-term resident of the city, he informed me that the Tank Stream had been Sydney's original river, that it was entirely covered over in the Victorian era and had been largely forgotten about ever since. Given my established fascination with the burial and subsequent ghostly presence of the Fleet within London's cityscape, the brief description was intriguing but – faced with the demands of establishing a home in a new city and commencing lecturing at a new university – I didn't have the opportunity to follow up on it for several years. My next connection was also an accidental one. During a walk around Circular Quay with my 4-year-old daughter, Rosa, in 1995, I paused while she played and splashed in a series of water features flowing down to the centre of the Quay (Fig. 2.3). I had walked past these on many previous occasions without having taken much notice of them. With time to spare, I looked around the lower feature

Fig. 2.3. Steven Walker's Tank Stream water feature, Circular Quay October 2023 (author's photograph).

and saw a descriptive plaque that identified it as 'The Children's Fountain' (1981) by Australian artist Stephen Walker and stated that was dedicated to 'the children who have played around the Tank Stream.' My curiousity was immediately piqued. The buried Tank Stream evidently had other memorials besides a bar name. Walking around the fountain, I saw second plaque that supplied an artist's statement that 'the Tank Stream fountain recalls mankind's past dependence on this flowing stream and our links with life around this region.' Following this up some days later I came across another statement from Walker, which specified that:

> My idea was to reintroduce flowing water back to this scene of the Tank Stream. I wanted to make a space for people to enjoy, with water and the suggestion of the flora and fauna along the Tank Stream. (*Canberra Times*, 1981, p. 8)

Some 15 years later, as my research turned to hydrology and urban landscapes, I started searching for historical references to the original stream through the National Archive of Australia's online Trove system. These searches revealed a continuing interest in the stream and its modern form as a central element of the city's sewer system. In 1934, for instance, a well-informed young correspondent named Rufus (identified as being 15 years of age), wrote a short item for *Sydney Morning Herald* that commented:

> I wonder if any of us, when going through the city, ever think of the Tank Stream, which perhaps is flowing under the street along which we are walking... Can you imagine a pretty, ferny valley and a stream with cabbage palms growing along its banks standing where Sydney, with its huge buildings, now stands?

Similar ideas were expressed more fancifully in Roderic Quinn's poem 'The Tank Stream', published in 1936, which opens with the protagonist slipping into a reverie and reconnecting to the pre-colonial past of the area where 'lay, twixt banks of Bracken/The Tanks Stream at my feet.'

Related themes were conveyed in a newspaper feature written by B. A. Broinowski in April 1945 that described a tour of the city's sewers and reflected on the past, present and future of the river. The article includes usefully detailed observations of the structure of the tunnels and the difficulties of following it:

> The going was slow owing to the slippery nature of the paved bed, which was mostly of stone, though in some places tiles were used. The guide said that these were Dalton Tiles which had been imported from England.
>
> The roof and walls were seen by the light of the torch to be of various materials, some portions being composed of "sandstock" bricks, which were hand made and of local manufacture about 90 years ago. Other portions are of stone or concrete. Here and there could be seen initials carved on the stones by the stone-mason gang of the early days and oozing from the walls the torches showed an iron deposit which the guide said came from the tramlines above.

The account concluded with the characteriation that:

> TAKING leave of this drab but still useful servant of the city, a vision arose. The city vanished and in its place a forest appeared a forest of mahogany, white gum, red gum, blood-wood, blackbutt and turpentine, she-oaks and wattles, Christmas bush and banksia. Here wildflowers lifted shining faces to the sun and only the wind and the birds broke the silence. Winding down a valley came a little stream, clear as crystal, finding its way by rocks and fern to the blue harbour.
>
> Such was the Tank Stream when Phillip saw it for the first time. Such was the stream on whose verdant banks he built the foundations of Sydney. (Broinowski, 1945, p. 8)

In a similar manner to Douglas's vision of an idyllic blue–green environment in Camden cued by the St Pancras water feature (discussed in Chapter 1), Broinowski conjured up an image of 'verdant banks' that sharply contrasted to both the subsurface sewer he had explored and the crowded city his account unfolds in.

While neither Trove nor the State Library of New South Wales's catalogue shows much evidence of public interest in the Tank Stream continuing in the decades immediately following World War Two, there is evidence of a significant rise interest at the end of the century. One notable public artwork, entitled *Tankstream – Into the head of the cove*, was commissioned from artist Lynne Roberts-Goodwin as part of the Sydney Sculpture Walk for the Sydney 2000 Olympics. The work comprised five small steel and glass motifs along the Tank Stream's former surface route through the city. The City of Sydney's website describes the installation in the following terms:

> The angles represented the flow of the Tank Stream in relation to the direction of Pitt Street – comparing the natural and man-made conduits of energy in the city. At the final site on Alfred Street the rods splayed to represent the delta as the stream flows into Sydney Cove. The movement of the subterranean stream was alluded to by the glass rods which were lit from below. They rippled blue light at night, creating a sense of flow beneath the surface.

Seeking to follow the artistic signposts over two decades after they were installed, I sought details online only to find that the artwork 'was deaccessioned in 2019' and had, thereby, slipped into as similarly unaccessible a history as the original stream that it had documented.

Coincidentally, 2019 was also the year in which the Tank Stream was first represented in a dedicated space. This location was far from incidental. In the 1850s, Sydney's first Post Office was deemed too small and insignificant to serve the needs of a rapidly expanding city and NSW architect James Barnett designed a suitably grand replacement using local sandstone. While the former PO building was demolished in 1863, construction proceeded slowly as it required the section of the Tank Stream running directly below the site to be securely enclosed. The new building opened to much fanfare in 1869 and operated for 120 years until it was closed and subsequently leased by the Sino Hotels group, extensively refurbished and opened as the Fullerton Hotel Sydney in 2019. As part of the renovations, the old sewerage conduit that ran underneath it was replaced by a steel pipe, with part of the old brick oviform sewer being preserved, along with 19th-century artefacts found in the sewers, as key elements in an interpretative display on the lower ground floor of the building. This remains the principal exhibit on the history of the buried stream and the materials used in its construction.

Access to and video documentation of the Tank Stream has been largely handled by the official agency responsible for the sewers, Sydney Water (SW), the latest version of a series of state water management bodies that commenced in 1888 when the NSW Board of Water Supply and Sewerage was established. Various incarnations of the Board persisted until 1994 when it was replaced by SW. In 2005 SW endorsed two associated documents, known collectively as the *Sydney Water Conservation Water Plan Manual* and *Tank Stream Conservation Management Plan* (henceforth referred to collectively as the CMP). The 'Executive Summary' of the documents identified the Tank Stream as important for state heritage since:

> The presence of the Tank Stream determined the placement of the settlement established by the arrival of the First Fleet in 1788, and its fresh water helped

> ensure the colony's survival. It is therefore intimately associated with the earliest moments of the European presence in Australia. As a result of that association, the Tank Stream has a symbolic value to the Australian community and this is reflected in a high recognition and interest in it as a heritage item. It is also important as an early stormwater channel, for its technological importance and other values, although these are less than the symbolic role. (Sydney Water, 2005, p. i)

Along with this statement, the Executive Summary also raised the issue of what the Tank Stream might be considered to comprise in the early 21st century:

> An issue identified in the CMP is that the physical item, its legal curtilage and the original course of the stream are all different. The original Tank Stream has been lost by later channelisation and development. The original channel has also been modified in sections through time. Little fabric from pre-1857, when it was channelised as part of the Sydney sewer and stormwater system, demonstrably survives, and the bulk of the older sections date from the mid- to late-nineteenth centuries, with extra sections that are less than fifty years old, as a result of underground basement construction.

The CMP also stressed that its references to the 'Tank Stream' encompass several meanings:

- The historical stream;
- The existing stormwater drain;
- The tunnel of brick and stone fabric enclosing the current storm water drain;
- The sections of the former route; and
- The idea of the tank stream, which has evolved over time and combines fact and fiction and is not necessarily related to reality. (SW, 2005 , p. i–ii).

This comprehensive definition of the stream is astute in acknowledging the various physical, temporal and conceptual planes that the Tank Stream now exists within (including 'those not necessarily connected to reality'). Fittingly, the CMP also identifies its limitations in being able 'to control activities outside the area directly under its own control,' most obviously, the 'idea' of the stream. (Sydney Water, 2005, p. ii). While it may not have been able to 'control' the 'idea' of the Tank Stream, it has played a prominent role in representing the buried stream to a contemporary public through producing and/or facilitating a number of videos of its route and structure.[4]

Before writing the conclusion to this chapter, I sought to reconnect with Sydney's historical streamlet by walking its surface route once again. I had hoped that a reconnection would inspire a rhetorical flourish that I could round this chapter off with, but the experience was less inspiring than I had hoped for. Walking from Hyde Park north towards the Fullerton Hotel and on to Tank Stream Way on a hot, late Spring morning proved a rather empty experience. The ghostliness of the river was – if anything – underlined by various

Fig. 2.4. Locked basement door leading to section of Tank Stream adjacent to Australia Square, November 2025 (author's photograph).

works underway in the city. Entering the Fullerton Hotel, I realized that major renovations were in progress in its lower ground floor and the stairs and escalator were roped off. I tried to persuade the staff to allow me a quick visit to the heritage display, but they were politely resolute in their refusal. Wandering up to Australia Square, past a barred basement doorway accompanied by a small plaque referencing the stream (Fig. 2.4), I walked to the fountain in Australia Square located above the old stream route and looked in vain for any commemorative plaque on it. This was disappointing, as I had previously ascertained that the fountain was designed by AW Roseman in 1986 to commemorate what was somewhat cryptically referred to as the '*construction* of the Old Tank Stream' (my emphasis) (City of Sydney, 2025), but it is debatable as to whether any observer might intuit a connection between a somewhat generic circular fountain and ghost river without signage. Moving northwards, I slipped through Tank Stream Way and passed the bar, which was closed. This northerly trudge was reinforcing a sense that settler society had definitively interned and buried the ancient stream and that the various placenames, obscure plaques and small memorials simply confirmed its pastness. I felt a sharp sadness and if I felt this way, how would any Gadigal walker trying to reconnect with the buried stream and their clan's precolonial past feel?

Weary and more than a little disheartened, I stopped, bought a coffee and headed into Angel Place where I lingered under Michael Hill's 'Forgotten Songs' installation. I've always liked the work. Visually, the 180 empty cages that hang airily above the laneway symbolize the many species of native birds that have been lost from the city over the last two centuries. Their songs play in the space all day before giving way to the calls of nocturnal birds that replace them in the evening. There is nothing in or around the installation (at least, that I have seen) that refers to the stream but I recalled that the installation's consultant ornithologist, Richard Major from the Australian Museum, had factored in the vegetation types that would have grown along the original stream when drawing up the list of bird species and songs featured in the final installation. Standing still with the birdsong and experiencing it as a sonic proxy of the original landscape at least offered me some connection to the stream's deep past and I took some solace from it. Heading on to complete the final short phase of my traverse, I visited Walker's (mercifully, still extant) Children's Fountain water feature before encountering a surprise. As I moved

Fig. 2.5. Hoarding boards around SW Tank Stream culvert restoration, Circular Quay, November 2025 (author's photograph).

towards the edge of the quay, intending to walk across the corner of First Fleet Park and over the culvertized stream's overflow point into the Harbour, my progress was arrested by a fence and hoarding boards (Fig. 2.5).

As I read on the boards, a SW restoration project was underway to ensure the culvert's structural integrity and functionality and maintain its 'cultural significance' in order to 'educate future generations about Sydney's early water management' (SW, 2024a). While the 'cultural significance' of the Tank Stream culvert, located close to Circular Quay, may primarily be understood in terms of its role in the city's early management of water flows, the various 'meanings' of the stream detailed in the previous discussion are, of course, complex. Each 'meaning' has its own significance. From SW's perspective, the instrumentality of the Tank Stream within the metropolitan sewer system it administers, and its desire to ensure that flood dangers are mitigated, are (understandably) dominant aspects. But what was most striking for me, was that SW had used the hoarding boards to produce a series of large-scale images of previous moments in the Tank Stream's history above the culvert that were more prominent than any other, longer-term memorialisation of the buried river elsewhere in the city. The hoarding displays were, thereby, a tantalisingly ephemeral recognition of the ghostly stream beneath the concrete that were soon to slip into a growing archive of lost acknowledgements.

As the attentive reader might be aware, I have avoided using the name 'Tank Stream' in the last few paragraphs as I have felt increasingly awkward about projecting a late 18th-century English Language term back into a pre-colonial past. The prosaic functionality of the term might be appropriate for its modern urban context but does not work so well in referring to a freshwater stream that once ran free across the landscape. Indeed, the unfortunate fact that its Indigenous name remained un-recorded by early British settlers compounds the indignity of its subterraneanisation. But, of course, there's another layer beneath that in that while rivers may have distinct identities, qualities and trajectories, their naming is a uniquely human activity. Placenames convey meanings to communities, and the loss of the stream's original name represents a loss of the associations that once accrued to it. The now nameless historic stream is thereby doubly effaced and irretrievable to the denizens of the modern city.

Notes

1 A now largely archaic term referring to an (ambiguously) small stream.

2 Once located around what is currently the intersection of Hamilton Street and Curtin Place, these appear to have been destroyed during the construction of the foundations of the multi-storey buildings erected at Australia Square in the 1960s.

3 See Townsend (1997) for a discussion of the picturesque aesthetic.

4 See, for instance, *The Tank Stream* (2013): https://www.youtube.com/watch?v=_ZKGSY8lJwM and *Experience the Tank Stream in VR 360* (2019): https://www.youtube.com/watch?v=gmyD45sAUdM

3

Ankyo Enthusiasm: Tokyo's Buried Rivers as Objects of Fascination

Junichi Hasegawa

3.1 Introduction

In Japan, in recent years, increasing attention has been drawn to individuals who are fascinated by *ankyo* – 'buried' natural rivers and artificial waterways, such as canals, that have been culvertized, infilled or covered over – and who have explored their traces in contemporary cityscapes. The term 'buried' emphasizes not only the changes to their physical infrastructures but also their historical memory, cultural meanings and perceptual dimensions within urban landscapes. This chapter examines the burial of rivers and waterways around Tokyo, the emergence of *ankyo* enthusiasts, their findings and assertions and the impact of hydrophilic projects including daylighting proposals. Among those *ankyo* enthusiasts, So Honda, Hideo Takayama and Nama Yoshimura are particularly noteworthy. While they originally began exploring buried rivers as a hobby, they have become recognized *ankyo* experts. Through their books,[1] social media activities, magazine articles, television appearances and public talks, they have led broader cultural engagement with these features. These three individuals each have their own blogs (Honda, n.d.; Takayama, n.d.; Yoshimura, n.d.) and Takayama and Yoshimura jointly run a blog titled *Ankyo Maniacs* which is also used as their collective name when they work together.[2]

The *ankyo* enthusiasts' activities have also been recognized by the academic community. They have contributed articles to journals published by the Japan Society of Civil Engineers and the Architectural Institute of Japan (Ankyo Maniacs, 2020, 2022), and have presented their research at a conference organized by the Japan Society of Hydrology and Water Resources and the Japanese Association of Hydrological Sciences (Shimomoto *et al.*, 2023). One of the co-presenters at the conference, Dr Shinichiro Nakamura, has reflected on the development and challenges of sociohydrology in Japan (Nakamura *et al.*, 2020) and organized the 2nd International Sociohydrology

Conference in Tokyo in 2025. The three researchers have also contributed to Mizbering, a public–private collaborative project initiated by the Ministry of Land, Infrastructure, Transport and Tourism (MLIT). This project aims to explore new possibilities for waterfront utilization to revitalize local communities by relaxing restrictions on riverbank occupancy and permitting land use such as commercial activities on riverbanks. The Mizbering website published a 12-part series on buried rivers, between September 2014 and December 2015. Three of these articles usually occupy all the top three spots in the website's 'popular posts' (Mizbering, 2026). Moreover, exploring buried rivers has become fashionable and chic. In a series in a men's fashion magazine (Miyazawa, 2021), popular young actor Hio Miyazawa, who describes himself as a fan of *ankyo* explorations, covered the theme in one episode, guided by Takayama. The *ankyo* enthusiasts have thereby transformed their personal hobby of exploring buried rivers into an academically grounded and fashionable social phenomenon.

In what follows, Section 2 provides an account of the extensive burial of rivers and canals in Tokyo during the 20th century. Section 3 situates the *ankyo enthusiasts* within a lineage of urban explorers and Section 4 describes the activities of Honda, Takayama and Yoshimura in more detail, drawing on their publications and blogs to show what kinds of rivers and sites they have focused on and how they have represented them. Section 5 has an auto-ethnographic focus, drawing on my experience in examining examples of canals and moats from the central and coastal sections of Chūō Ward. Section 6 considers the influence of hydrophilic activities on the establishment of waterside parks and daylighting attempts in connection with the restoration of Shibuyagawa[3] in the vicinity of Shibuya Station and the relocation of an expressway underground in the Nihonbashi area. Finally, the concluding section briefly summarizes the discussion.

3.2 The Triggers and Processes of Burying Rivers and Their Influence on the Perception of Them

Historically, the major developments that prompted the emergence of buried rivers were the post-1923 Great Kantō Earthquake reconstruction, the post-1945 Second World War war-damage reconstruction, the construction of expressways and the development of sewerage systems during the period of rapid economic growth that began in the late 1950s (Figs. 3.1 and 3.2). In both post-earthquake and post-war reconstruction, river and canal infill was undertaken as a means of disposing of debris from damaged buildings and other structures. During the construction of the Metropolitan Expressway, which was accelerated to meet the deadline of the 1964 Tokyo Olympics, a policy was adopted to make maximum use of river and canal routes in order to minimize the need for land acquisition from private owners. Although some waterways – such as the Tsukijigawa – were reclaimed through riverbed landfill, others, including the Kandagawa, Nihonbashigawa, Furukawa and Sotobori outer moat, were left as waterways but had elevated expressways built above them, effectively obscuring them from view.

Sewerage development, meanwhile, was undertaken based on the so-called 36th Report on the rivers and sewerage submitted to the Tokyo governor in 1961. Under this plan covering 14 rivers (11 in the western wards – including Megurogawa, Kitazawagawa, Karasuyamagawa, Jakuzuregawa, Furukawa, Shibuyagawa and Momozonogawa – and 3 in the eastern ward of Edogawa Ward), 6 were legally abolished as rivers and converted into sewer trunk lines or covered over; 7 remained subject to dual management under the River Act and Sewerage Act (and only the Furukawa retained its status as a statutory river) (Ishihara *et al.*, 2019). In this way, the favourite sites of the *ankyo* enthusiasts gradually took form.

Under the reconstruction plan formulated in the aftermath of the Great Kantō Earthquake, three canals located in what is now the eastern part of Tokyo's central wards – Irifunegawa, Teppozugawa and Nishihoridomegawa – were filled in using debris generated by the earthquake. Larger-scale landfill projects for the disposal of debris, however, were undertaken through coastal reclamation, including the creation of the area that now hosts the Toyosu Wholesale Market. Furthermore, Tokyo City – which subsequently became the Tokyo Metropolitan Government following its 1943 merger with Tokyo Prefecture – petitioned the national government for the construction and improvement of canals.[4] The City of Tokyo's initial request for the extension

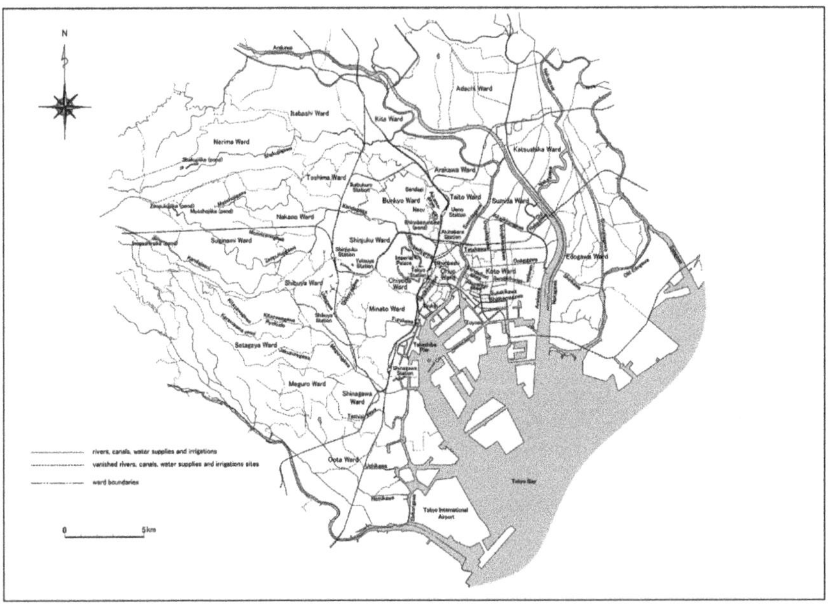

Fig. 3.1. Map of rivers and canals in Tokyo's ward areas (author's design).

Fig. 3.2. Map of rivers and canals in Tokyo's central and eastern ward areas that have been buried (indicated by dashed lines) (author's design).

of either Higashihoridomegawa or Nishihoridomegawa to Kandagawa was, however, rejected by the government because plans for new canal construction and renovation were being scaled back due to reconstruction budget cuts. Consequently, a decision was made in accordance with the request from the local ward assembly that, if reaching Kandagawa was not possible,

Nishihoridomnegawa should be infilled because the dead-end canal might pose a threat to sanitation (Shoji, 1989; Hasegawa, 2023, pp. 96–98).

Compared with the relatively limited infilling undertaken as part of the reconstruction planning following the Great Kantō Earthquake, the post-war war-damage reconstruction plan proposed the infilling of a greater number of waterways. This targeted the canals located in the central and eastern parts of the Tokyo wards, whereas rivers in the western districts were slated primarily for improvement rather than burial. In 1947, 14 canals totalling 6.762 km were designated as obsolete waterways to be infilled as they were no longer deemed to fulfil their original function. In the following year, the decision was made to infill the 1.108-km-long Sanjikkenbori canal, which then marked the eastern edge of Ginza, Japan's foremost commercial and entertainment district. This policy was partly driven by a severe backlog in disposal of rubble resulting from wartime air raids. Among these waterways, *ankyo* enthusiasts have presented the Ryūkangawa, Gokenborigawa and Rokkenborigawa canals (located in the Kōtō and Sumida wards – the eastern part of Tokyo's ward area) in a highly positive manner (e.g. Yoshimura, 2015, 2020a; Honda, 2023). Relatively little attention, however, has been paid by *ankyo* enthusiasts to the fact that the development carried out on land created by filling in the Sanjikkenbori canal with wartime rubble caused a major social controversy (Yoshimura, 2020b, pp. 76–79). The Tokyo Metropolitan Government, which had overseen the reclamation, had initially expressed its intention to promote cultural- and public-oriented development. Nevertheless, the first substantial building constructed there in 1951 was an entertainment complex named Tokyo Onsen (*Tokyo Spa*), consisting of sexually suggestive sauna facilities and high-end bars and restaurants that catered mainly to corporate clientele. Coming just 1 year before Japan's restoration of sovereignty, this development was widely criticized in newspapers and magazines of the time, and ultimately compelled the prime minister to pledge appropriate action during a session of the National Diet.

Around the time Tokyo Onsen opened, the Tokyo Metropolitan Government sought to advance a plan in which the Tokyo Expressway Company would construct a row of two-storey buildings running 1.3 km along Sotobori's outer moat, with its rooftop being used as a roadway. Suspicions soon arose that the metropolitan government's true objective was not the construction of an elevated motorway but rather the creation of a major, two-storey-high shopping complex. The issue was widely discussed in the media and in the National Diet, with reports also highlighting opposition from local business associations. Indeed, although the roadway portion did not open until 1959, the buildings beneath it had already become a thriving shopping centre in the years prior to the road's completion (Hasegawa, 2015a, b, 2018, 2023).

The Tokyo Expressway is physically integrated with the Metropolitan Expressway, which opened later and formed an extensive expressway network in the city. The process leading up to the decision to construct its initial eight routes unfolded roughly as follows. In Tokyo's post-war reconstruction planning, a wide arterial road network had been proposed. However, due to financial constraints road development progressed slowly, and severe congestion emerged as enormous volumes of intra-city traffic accumulated. In 1957, the Ministry of Construction formulated a basic selection policy for the expressway's routes that unused land and rivers or canals that posed no problems for flood control or water use should be utilized wherever possible. In response, a special committee for expressway investigation was established within the Tokyo Metropolitan City Planning Local Council in the same year. Based on this committee's recommendations, in August 1959 the Minister of Construction officially announced an initial plan for eight routes. Of the 62.5 km total length of the initial eight routes of the Metropolitan Expressway, 41% was constructed above rivers, compared with 37% over streets and 22% over ordinary residential land, in accordance with the Ministry of Construction's 1957 policy. Even at Nihonbashi[5] – a bridge on the Nihonbashigawa canal of historical significance as the point of origin of Japan's nationwide road network in the Edo period – the waterway was covered by an expressway running atop piers built in the river.

The sewerage development program was allocated funding as part of the 1964 Olympic projects, which accelerated its implementation. As early as 1950, Tokyo's sewerage plan had articulated a policy whereby rivers in the metropolitan wards whose sources had dried up were to be converted into main sewer conduits. In combination with the Ministry of Construction's 1957 policy on urban expressways and the 36th Report on the rivers and sewerages, both noted above, and the pressing need to expedite preparations for the Olympic Games, this policy of conversion into main sewer conduits further propelled the culverting of rivers. Recent scholarship, together with testimonies from individuals involved at the time, underscores that the rivers had become severely polluted due to illegal dumping, delays in sewerage development, and the rapid spread of synthetic detergents was widely perceived by residents as a public nuisance. Consequently, in the formulation and the implementation of sewerage development or urban expressway construction policies, discussions proceeded with the prevailing assumption that the repurposing of rivers for sewerage functions was both desirable and unavoidable (Ishihara and Takasaki, 2015; Ishihara *et al.*, 2015; 2018; Magata, 1996; Nakamura and Oki, 2009).

As shown in this section, the burial of rivers was carried out on a far larger scale during the rapid economic growth of the 1950s – particularly in connection with expressway construction and sewerage development – than during the post–Great Kantō Earthquake or in the post–World War reconstruction periods. Many of the rivers affected during this phase are precisely those that have attracted the interest of *ankyo* enthusiasts. By contrast, both the development of the Sanjikkenbori canal area, which was reclaimed during post-war reconstruction, and the construction of the Tokyo Expressway, built in the 1950s by filling in the Sotobori outer moat, came to be perceived as scandalous and driven by profiteering, and consequently acquired a negative public image. In these projects, it was difficult to discern any sense of pride on the part of individuals or local communities.

3.3 The Backgrounds of *Ankyo* Enthusiasts

The activities of *ankyo* enthusiasts can be seen to relate to the development of previous activities that involved walking through the city and conducting fieldwork-style observation. Yoshimura refers to *rojō kansatsu* (street observation), a form of urban walking popularized in the 1980s by avant-garde artist and writer Genpei Akasegawa, architectural historian and architect Terunobu Fujimori and others as a precedent. *Rojō kansatsu* enthusiasts, in turn, trace their origins to *kōgengaku* ('modernology') – an approach proposed in the late 1920s by architectural scholar and folklorist Wajirō Kon in the aftermath of the Great Kantō Earthquake. This sought to analyse and interpret contemporary social phenomena through the systematic observation and study of specific places and times. Drawing on this, *rojō kansatsu* enthusiasts observed and documented overlooked elements of the urban environment – such as ordinary buildings, signboards, manholes and other features – and published a book on the topic in 1986 (Akasegawa *et al.*, 1986). Their method of observation, which often explored the city from a humorous perspective, soon became a cultural trend (Sawayama, n.d.).[6]

Increasing numbers of people have become interested in exploring a variety of hidden urban topographies and remains, such as slopes, cliffs, stairways and stone walls along with buried rivers. This trend originated in the 1980s with movements as street observation, as mentioned above. During the early-mid 1990s, following the collapse of Japan's economic bubble, *sanpo* (walking or strolling) gained popularity as a low-cost way to enjoy familiar urban environments during a recession. Urban walks aimed at enjoying local attractions and cuisine developed into a distinct market and *Sanpo no tatsujin* ('Master Walker') – a magazine specializing in walking – was launched in 1996. In the 2000s, more people began to enjoy *machi-aruki* ('city walking') beyond conventional tourist routes and television programs such as *Chii Sanpo* ('Chii's walking') launched in 2006, gained popularity. This program featured the late actor Takeo Chii as the main host who strolled through various places himself. It continues today, with its third host, actor and comedian Junji Takada, and with its title having changed to *Jun Sanpo* (Jun[ji]'s walking).

The NHK (*Nippon Hoso Kyokai* - Japan Broadcasting Corporation) program *Buratamori*, which started in 2008 and was serialized in 2009, revolutionized this trend by introducing a new form of city walking centred on learning about the city. The program allowed the viewing public to take pleasure in observing enthusiasts' delight in their own obsessions, leading to an increase in invitations to leading *ankyo* enthusiasts for talks and other events.[7] In the program, Tamori, the celebrity host, strolls through various neighbourhoods and uncovers the origins and development of the city by interpreting its topography and landscape. Rather than merely introducing current shops and streetscapes, *Buratamori* deciphers the hidden topography and historical structures beneath them by interpreting various traces and then uses computer graphics to recreate historical landscapes that differ from what we see today. It has turned such acts of interpretation into a form of entertainment (Takenouchi, 2015; Takayama, 2025). The Tokyo Suribachi Society has also contributed to the expansion of this kind of urban 'excavation' tourism. The society was formed in 2003 by architect Norihisa Minagawa, who works for Kajima Corporation, a major construction company. The society observes and documents Tokyo's *suribachi* topography, i.e. bowl-shaped depressions in the cityscape.[8] It has organized numerous fieldwork excursions and lectures and Minagawa has appeared on many occasions on programs such as *Buratamori*, and Honda, Yoshimura and Takayama have all contributed columns and essays to Minagawa's publications (Takenouchi, 2015).

As we have seen in this section, *ankyo* enthusiasts emerged from Japan's tradition of street observation rooted in *kōgengaku* and 1980s urban walking movements. Interest expanded through 1990s' strolling culture, specialized magazines, and televized *machi-aruki*. Programs such as *Buratamori* popularized interpretative urban exploration as entertainment, while groups like the Tokyo *Suribachi* Society further advanced field-based engagement with hidden topographies and urban 'excavation' tourism. In the next section, we will examine the respective arguments and philosophies of Honda, Yoshimura and Takayama, who became *ankyo* enthusiasts through auto-ethnographic documentation of personal experiences and encounters.

3.4 *Ankyo* Enthusiasts

The three *ankyo* enthusiasts focused on in this chapter – Honda, Takayama and Yoshimura – each became fascinated with buried rivers through personal experiences and began exploring buried rivers as a hobby. Born in 1972, Honda first realized that a river had once flowed through the neighborhood where he lived when he obtained an old subdivision map of Tokyo as an elementary school boy. This realization marked the starting point of his search for lost and invisible rivers and spring sites. Since 1997, he has been publishing the results of his explorations on his own website. For him, exploring *ankyo* is an act of 'seeking the memory of [the] original landscape that had already been lost.' By tracing the former courses of rivers and the vanished hydro-geomorphological features, he attempts to reconnect the 'memory of the land surrounding water' with the present urban space (Honda, 2024).

Honda systematically interprets *ankyo* through three axes – space, time and landscape. That is, he views them in terms of the temporal layers of the river's life cycle: the period when the river was open to the air, the period when it was converted into a culvert and the period after its burial. He also focuses on landscape clues that remain above ground, such as bridges, remnants of embankments, winding alleys, retaining walls and exposed drainage pipes. Furthermore, he considers their spatial dimensions, including location, topography, elevation differences and networks of irrigation canals. He uses the term *ankyo-scape* to refer collectively to these distinctive landscapes, terrains and traces of buried rivers. These include elements from the pre-burial period – such as bridge pedestals, retaining walls, former dyeing workshops, fishing ponds, pools and meandering paths – and those from the post-burial present, such as concrete lids, exposed drainage pipes, greenways and bollards. Exploring *ankyo* is thus an act of rediscovering vanished waterways – and, by extension, the face of the city itself. It also serves as a means of interpreting the complex interwoven processes underlying their burial: urbanization, conversion into sewer systems, earthquake damage, war destruction and residential land development (Honda, 2024).

Takayama recalls that he first encountered *ankyo* in the 2000s in Setagaya ward where he had lived most of the time since his university studies in 1983. He recalls that the encounter made him aware of 'the *ankyo* within his own heart' and remarks that 'everyone carries an *ankyo* inside them.' For him *ankyo* are small, damp and quietly decaying ruins that are unnoticed and unappreciated. Yet even such *ankyo* retain traces of their former water surfaces. They quietly preserve the pride of 'once having been a river.' No matter how they diminish, they teach us that one's unique dignity can never be lost (Takayama, 2022). His discourse can be interpreted as a form of metaphorical personification of urban ruins. He also identifies three elements that constitute the appeal of *ankyo*. First, by tracing the routes of former waterways, one can rediscover the invisible and hidden networks of water. Second, by investigating how these rivers came to be buried, the unique historical layers of each locality are revealed. Third, by encountering *ankyo signs* – such as remnants of riverbanks or bollards – that indicate where rivers once flowed, one can appreciate their aesthetic charm, akin to that of antiques or Japanese gardens (Takayama, 2016). Among these, Takayama places particular emphasis on bridge remains, which he calls *ankyou* (literally 'dark bridges') – the most reliable landmarks of an *ankyo*. In particular, he values *nora-ankyou* (literally 'wild dark bridges') – those so forgotten that no one even recognizes them as former bridge sites – arguing that discovering such places becomes a source of pride both for the discoverer and the neighborhood itself (Takayama, 2025).

In the late 2000s, Yoshimura discovered that a greenway running through a shopping street in Kōenji, Suginami Ward – a residential district in western Tokyo that she had often visited since her student days – had once been the Momozonogawa, now buried underground. From that moment, she says, she 'fell in love with *ankyo* for the first time' and became captivated by their charm (Yoshimura and Takayama, 2021, p. 3). As a 'deep-digging *ankyo* researcher,' she combines local histories, maps and archival records to elucidate in detail the background of each river's burial (Yoshimura, 2025). At the same time, she emphasizes the idea of *ankyo* as a connector. Defined as 'spaces that were once rivers or waterways,' *ankyo* link places within the same watershed, allowing them 'to connect instantly, like synaptic transmission' and to evoke in people of those areas a sense of kinship, as though the towns along the same water system were 'sister cities.' At the same time, *ankyo* also connect lands that were once divided by waterways. Once a river becomes an *ankyo*, the barrier disappears. In this sense, the very act of becoming buried – a process of *ankyo-ka* – is itself an act of connection (Yoshimura, 2025).

In the major works produced by *ankyo* enthusiasts, particular attention is devoted to the rivers in western Tokyo's 23 wards. These waterways were essentially natural rivers or agricultural channels through which water once flowed according to variations in elevation. They were situated in such metropolitan wards as Setagaya, Meguro, Shibuya, Suginami and Bunkyō, where distinct topographic contrasts existed between upland plateaus and the valleys carved by the erosion of these hills – valleys characterized by narrow floors and gently sloping sides – and where springs, escarpments and tributary streams once shaped the landscape.[9] By contrast, the buried rivers of the eastern part of Tokyo's 23 wards – such as Chūō and Kōtō – were fundamentally canals and moats constructed in the early Edo period in the 17th century. These waterways were artificial structures built for the defence of Edo Castle and for transportation, located in coastal lowlands created through land reclamation, areas at or near zero metres above sea level with virtually no topographic variation and originally without natural water flow. Signs of *ankyo* features that *ankyo* enthusiasts typically value are relatively scarce in these districts. Consequently, although these eastern waterways have been addressed and have recently begun to receive more serious attention by *ankyo* enthusiasts, their treatment has remained relatively limited. Moreover, the differing degrees of attention that *ankyo* enthusiasts have paid to particular buried rivers appear to be related to the timing, purposes and characteristics of their culverting, infilling or covering of the space above them.

3.5 A Personal Perspective

I was born in 1961 and lived in Nihonbashi-Muromachi. As an infant, I would be taken for daily strolls to Nihonbashi bridge in my baby carriage (Fig. 3.3), and I often went to the Mitsukoshi department store located right next to the bridge. I was, apparently, something of a favourite among the female sales clerks there, although I have no memory of this, nor have I had a similar experience since. Without an elevated roadway overhead (Fig. 3.3), Nihonbashi was apparently a pleasant place, at least in terms of its townscape. However, air pollution in the city centre was severe in the early 1960s, since automobile engine emission regulations were only introduced in Japan later in the decade. As a result, I was frequently ill and my family moved to the western suburbs of Tokyo to ameliorate the problem. When I was ten years old, we eventually settled in Suginami Ward, where the open-channel Zenpukujigawa flows through the neighborhood. From the 2000s onward, while researching

Fig. 3.3. The author in a baby carriage in Nihonbashi in 1962. (Courtesy of Ryoko Hasegawa.)

Tokyo's post-war reconstruction, I became particularly interested in the infilling of Sanjikkenbori and the Sotobori outer moat, and in the development carried out on the reclaimed land created through these processes. My primary concern was academic in nature: I approached buried rivers as political and social problems – as products of flawed urban policy. The central aim of this research was to situate this case within the broader context of post-war reconstruction and urban history.

In examining post-infilling development at Sanjikkenbori, I analyzed the Miharabashi underground shopping mall, a development that has attracted the attention of *ankyo* enthusiasts. In this instance as well, the media criticized the nature of the commercial area, which was lined with cheap *izakaya* bars and *pachinko* parlours (then widely regarded as a social problem). In 1954, the deputy governor of Tokyo was compelled to offer an apology for the development before the National Diet (Hasegawa, 2015a, pp. 495–496; Hasegawa, 2023, pp. 109–110). At the same time, this underground shopping mall, built beneath the Miharabashi bridge, did indeed contain what would now be described as *ankyo* signs or *ankyou* – dark bridges. When I photographed the mall in 2013, I was unfamiliar with the term *ankyo*. Nevertheless, the mall's gently curved ceiling clearly indicated that the structure had once been a bridge, powerfully asserting its former presence. The atmosphere of the shopping mall – composed of inexpensive eateries and a small cinema – also felt dated. I found myself drawn in and took photographs almost instinctively

Fig. 3.4. The author (right, with back to camera) stands in a similar position to Fig. 3.3 in Nihonbashi in 2025. (Courtesy of Ayaka and Sanshiro Hasegawa.)

Fig. 3.5. Miharabashi underground mall in 2013 (author's photo).

(Fig. 3.5). The underground mall was, however, closed the year after I took those photographs, and it has since been filled in. Today, there remain few connectors in the vicinity of Miharabashi to its past as a waterway.

3.6 Recent Developments in Relation to Daylighting Initiatives

The view that rivers should be regarded merely as de facto land for sewer infrastructure or expressway construction came under criticism in the late 20th century and began to change. As evidenced by the evolution of three river improvement plans in Kōtō Ward, approaches to environmental planning shifted significantly over time. While the 1971 plan called for filling in and culverting unnecessary waterways to increase the efficiency of land use, the 1989 plan reconsidered this approach and emphasized the importance of incorporating waterside amenities, reflecting a renewed appreciation for the value of waterfronts and water surfaces. The 1997 plan included proposals for widening channels and integrating waterways with the surrounding townscape (Inomata *et al.*, 2000, pp. 188–189).

Since the 1970s, numerous waterside parks (*shin-sui kōen*) have been developed to create spaces where people can enjoy proximity to water. In the case of culverted rivers, the areas above them were often converted into greenways (*ryokudō*) or equipped with artificial water features. However, some critics argue that many of these parks offer only superficial simulations of natural water flow and nature, amounting to little more than cosmetically fabricated 'waterside' environments (Nakamura and Oki, 2009, p. 570).

In recent years, plans to restore the Shibuyagawa in the vicinity of Shibuya Station have attracted considerable attention for measures such as relocating the discharge point of treated wastewater further upstream, installing waterscape facilities that spray this treated water from the top of the embankment, and developing riverside promenades. These initiatives have been regarded as potential precursors to the daylighting of the culverted sections. However, the primary purpose of this restoration plan is for the developers to undertake what appear as environmental 'improvements,' in anticipation of development incentives such as floor-area ratio bonuses (Ishihara *et al.*, 2017, pp. I_298–I_299). Whether waterside restoration carried out for large-scale, commercially oriented real estate developments can truly be considered a revival of the river's former clear stream is hardly convincing in the light of the nature of the history of recent large redevelopment projects in Tokyo (Hasegawa, 2025, 2026).

Regarding the use of river corridors as routes for expressways, growing criticism – particularly on aesthetic and environmental grounds – led the Ministry of Construction's River Bureau in 1973 to prohibit the installation of bridge piers and similar structures within river channels (Ishihara and Takasaki, 2015, pp. 94–95). In the case of Nihonbashi, discussions on relocating the expressway underground began in the early 21st century. Although the process was not without complications, construction finally commenced in 2025. Nevertheless, the segment to be relocated underground amounts to only about 1.8 km of an expressway network that now exceeds 300 km in total length. Even so, the removal of the elevated section above Nihonbashi is not expected until 2040, and the project is a major undertaking with an estimated cost of 320 billion yen (Ministry of Land, Infrastructure, Transport and Tourism, 2018; Metropolitan Expressway Company Limited, n.d.). While this project can be considered part of five large-scale redevelopment initiatives undertaken by major developers and others (Nihonbashi River Walk, 2025), these efforts cannot be taken at face value and require critical evaluation.

3.7 Conclusion

In this chapter I have examined the background to the emergence of *ankyo* enthusiasts, their contributions to raising public awareness of buried waterways, the ways in which patterns of river burial have shaped their interests and recent attempts to consider the daylighting stretches of buried rivers. As walking and urban strolling gained media attention as leisure activities – most notably through the growing popularity of *Buratamori*, which revealed hidden urban histories – *ankyo* enthusiasts emerged and began to engage actively in exploration. They placed particular emphasis on traces of former rivers and on topographies marked by changes in elevation, rediscovering past waterways and hydrological systems. In doing so, they conceive of buried rivers as connectors across time and space, through which they find personal and local pride. Whether a given site evokes a sense of such pride and empathy seems to be the decisive factor in determining its significance for them. By contrast, their interest is extremely limited in cases such as the Sanjikkenbori canal or the Tokyo Expressway on reclaimed Sotobori outer moat, where they find that such feelings of pride and empathy are difficult to elicit. Contemporary daylighting initiatives likewise do not appear to generate these same connections to pride or empathy. They have their own distinct approaches to reviving buried waterways, symbolized by their practice of tracing the remnants of old waterways and bridges. If implemented, daylighting based on large-scale civil engineering projects and urban redevelopment would probably be appreciated by Tokyo's inhabitants, but it is not the primary focus of *ankyo* enthusiasts. Finally, while buried rivers have rarely evoked fear or been perceived to pose direct threats to Tokyo, extreme rainfall associated with climate change has led to an increase in flows that exceed sewer system capacity (Ministry of Land, Infrastructure, Transport and Tourism, n.d.). While the Metropolitan Area Outer Underground Discharge Channel (completed in 2006 and currently being upgraded) has helped the city's overall flood resilience, there is a danger of buried rivers bursting their confines and returning to the surface of central Tokyo during peak precipitation events with disastrous consequences; and their current structures and management thereby require urgent consideration.

Notes

[1] Particularly, Honda, 2017, 2023; Honda (ed.) 2012, 2021; Honda *et al.*, 2017; Takayama and Yoshimura, 2020, 2023; Yoshimura and Takayama, 2015, 2021, 2024

[2] Details of their activities are found in 'Works' section of the Ankyo Maniacs blog and in the 'Books and Magazines' and 'Events' sections of Honda's blog.

[3] The suffix *-gawa* indicates a river but it is also frequently used for canals.

[4] As part of this initiative, it sought to extend either the Higashihoridomegawa or the Nishihoridomegawa to connect with the Kandagawa, thereby enabling waterborne transport between the freight stations at Shiodome and Akihabara. For short-distance, small-scale freight distribution within the city, the canal network continued to play an important role, owing largely to the availability of low-wage labour.

[5] The suffix *-bashi* denotes a bridge.

[6] Although Yoshimura notes that she had been unaware of the 1980s' boom at the time and only became interested in it after becoming absorbed in buried rivers, treating it as a related genre. (Yoshimura and Takayama, 2021, p. 3).

[7] In a fascinating English language blog about exploring Tokyo, Yoshimura and Takayama (2015) is cited, and the blog administrator introduces numerous buried rivers (Parker, 2016).

[8] The society takes its name from a *suribachi*, type of ceramic grinding bowl.

[9] The rivers most prominently emphasized in these works include Shibuyagawa and its tributary Udagawa; Megurogawa and its tributaries such as Jakuzuregawa, Kitazawagawa and Karasuyamagawa; Tsurumakigawa; Yabatagawa, which flowed through Itabashi and Kita Wards in north-western Tokyo; the tributary systems of Kandagawa –such as Myōshōjigawa and Zenpukujigawa; Momozonogawa; the former waterways and bridge sites around the Sotobori outer moat of Edo Castle and Akasaka-mitsuke and Aizomegawa, which once flowed through Nezu and Sendagi – a popular walking site of traditional Tokyo residential area.

4 Cheonggyecheon, Daylighting Projects and Urban Transformation

4.1 Introduction

Having considered various aspects of the persistence of awareness of buried rivers in urban landscapes and a range of socio-cultural acknowledgements and elaborations of this, we now turn to the practice of daylighting and its motivations. As emphasized in the Introduction to the volume, there is no necessary or direct connection between abiding public fascination with buried rivers of the types referred to in Chapters 1–3 and the successful proposal and acquittal of daylighting projects. Indeed, with regard to the Fleet, Tank Stream and Tokyo's buried rivers and canals, it is notable that these have not been subject to the daylighting that has occurred in a number of other locations. Indeed – and conversely – one factor shared by the principal urban daylighting projects undertaken to date is that their rivers were not subject to sustained socio-cultural recollection and re-imagination before their deculvertization.

Wild, Dempsey and Broadhead (2019) have provided insightful research into a range of 'drivers' for daylighting projects reported by international participants in a web-based survey that covered 90 deculverting projects. The most prominent factor they identified was ecological/habitat related (32 examples), followed by flood prevention (23), development/redevelopment (13), creation of an amenity/park/greenspace (12), a long tail that included aesthetic/artistic/landscape factors (4) and, in equal last place, culture (1) (2019, pp. 257–258). History was not reported as a factor. Given the prominence of landscape, culture and history in the perceptions and writings of English, Australian and Japanese ghost river/*ankyo* enthusiasts discussed in Chapters 1–3, the disjuncture between their focus and the data reported by Wild, Dempsey and Broadhead is clear.

Histories of river daylighting often commence with the Strawberry Creek project conducted on the Berkeley campus of the University of California in mid-1980s and there are several insightful published accounts of its history and process.[1] However, we do not attend to these in this chapter since the deculverting aspect of the project concerned a small, covered-over stretch of a creek that ran through a relatively low-density university campus. Nor do we address the daylighting of small streams in Zurich in the late 1980s and early 1990s undertaken in implementation of the city's *Bachkonzept*.[2] While the latter project is notable for having extricated buried streams from the city's sewerage network and restored surface flows to parks, former laneways and private gardens, it generated these as low-key environmental features intended to enhance neighbourhoods without major modification of the communities concerned. Our focus in this volume is on more transformative projects.

4.2 Kalamazoo

The first notable (albeit often unheralded) urban daylighting project was undertaken on Arcadia Creek, in Kalamazoo, Michigan (USA) in the mid-1990s. In the early-mid 1800s, Arcadia Creek was subject to intense pollution and was culvertized between 1873 and 1890 to contain this. Hoobyar (2002) has identified that the daylighting of the river that occurred over one hundred years later did not result from any attempt to recreate a pre-culverted version of the creek (let alone recognition of and engagement with First Nations communities' prior usage and stewardship of it) but rather from a more functional project to revitalize the city's north-central business district and tackle frequent flooding in the area. The project was a top-down initiative from the city commission and the daylighted area consisted of a 1.2 km concrete channel with a parallel walking track and an open, 3.5-m-deep stormwater pond that could double as a sunken amphitheatre, creating what Hoobyar (2002) refers to as 'an openair, unfenced oasis in the urban core'. Despite the project's disconnection from the river's past, the engineered section of Arcadia Creek has been popular as a public space and its amphitheatre proved successful as a festival venue that attracted tourists and related business opportunities to the area in the decade following its completion.

One notable aspect of 'first wave' urban river daylighting projects, such as Arcadia Creek, is that their design was premised on perceptions of predictable variations in river height and intensity of flow, precipitation and broader weather patterns over the immediately preceding decades. While this no doubt appeared reasonable to its designers, such calculations have become particularly problematic over the last 30–40 years given the increasing impact of global warming and related climate change. One effect of this in locations such as Michigan has been an increased intensity of storms (Center for Climate Change Integrity, 2025) and, thereby, risks of daylighted rivers flooding their banks as a result of peak downstream flow and/or substantial runoff from surrounding hard surfaces. In acknowledgement of this, in August 2025, the City of Kalamazoo announced a building works program to improve the capacity of existing culverts and to provide a substantial upgrade and extension of the daylighted area of Arcadia Creek in order 'to handle the intense and more frequent storms Kalamazoo is experiencing due to climate change'. The official announcement also notably situated the event in an historical context:

> This project builds on past work to improve Arcadia Creek, including similar floodplain removal efforts completed downtown in the early 1990s. With this next chapter, Kalamazoo is taking a bold step toward a more resilient, sustainable future. By rebuilding Arcadia Creek, we are not only addressing urgent infrastructure needs, we're also laying the foundation for environmental restoration, economic development, and stronger connections between our neighborhoods and downtown.

This recognition of the need for redesign of the daylighted riverscape and its surroundings to meet climate change, and the opportunity to capitalize on such developments for civic renewal, emphasizes that many (if not all) urban daylighting projects realized to date are likely to require modifications in response to changing global climate conditions.

While Kalamazoo was a significant pioneer of urban river daylighting – and is now pioneering climate adaptation strategies – the project that has received most international attention is Seoul's Cheonggyecheon stream, the most extensive and most transformative urban daylighting venture undertaken to date.

4.3 Cheonggyecheon

The Cheonggyecheon daylighting project occurred along a formerly in-filled stream in Seoul in 2003–2005. The stream is relatively short, running 14 km west from its source through Seoul before joining up with the Jungnangcheon tributary of the Han River. Like the Fleet, discussed in Chapter 1, Cheonggyecheon had been subject to a variety of interventions prior to its downtown stretch being covered over in 1951–1958 and had a complex sociocultural history before it was subjected to a major redesign in the early 2000s. Reflecting both the scale of Seoul's project and substantial international interest in the redevelopment, Kim and Jung's systematic literature survey of the project (2019) identified 86 studies of it published between 2003 and 2017, and further material has followed over the last decade.

Seoul became the capital of Korea in the 1390s when the emergent Joseon dynasty decided to establish a new city on the banks of the Cheonggyecheon. One issue with the new capital site was that the stream often spilled over onto its floodplain during the summer rainy season. Following a particularly bad flood in 1410, a number of engineering works were carried out. The stream was widened, deepened and lined with embankments, and tributary flows were directed away from it to avoid flash flooding. Kim and Jung (2019, p. 14) have identified this initiative as 'a landmark river engineering project' involving the recruitment of 200,000 people. As the city grew, there were predictable contrasting pressures on the waterway. Originally used as a fresh water source, it became badly polluted with sewerage. Indeed, when Japan occupied

Korea in 1910 the stream's heavily polluted nature was seen as a significant problem. Extensive dredging work was undertaken in 1918 to improve flow and several areas were covered over but the stream remained in a dire state for the remainder of the Japanese occupation period (which concluded in 1945) (Fig. 4.1).

Following the Korean War (1950–1953), which resulted in the division of the peninsula, South Korea underwent a period of rapid industrialization, particularly around Seoul, whose population grew as migrants from rural areas arrived to work in new factories and service industries. With pressure on available land in the city, the area around the heavily polluted and increasingly silted-up Cheonggyecheon, which had become typified by low-quality houses and narrow and increasingly traffic-choked streets, was targeted for redevelopment. One result was the covering over of stretches of the stream's course between 1958 and 1977, which allowed various forms of marginal economic activity such as 'flea markets'[3] to develop in the newly available spaces. In 1976, the area was further transformed by the completion of an elevated highway running over the stream's former route that became one of the capital's busiest roads and a symbol of the city's economic boom. Perceptions of the area at this time vary markedly. Yoon (2022, p. 14), for instance, has contended that there was 'a very vibrant community, with social and political complexities' located underneath the expressway. This community occupied a network of small plots and traditional narrow laneways that also housed small-scale metalwork and manufacturing operations, while 'lower-income workers, merchants, customers and illegal vendors populated the downtown area every day and built their livelihoods around it' (Yoon, 2022, p. 14). The

Fig. 4.1. A stretch of the Cheonggyecheon with overhanging dwellings in 1945 (Wikimedia Commons, 2009).

perspectives of many external observers were significantly different, with the locale, occupants and livelihood activities being seen as out-of-step with a rapidly modernising city whose leaders were trying to project an image of vibrant modernity. One factor that brought these disagreements to a head was the deterioration of the highway constructed over the stream. Sceptical about the value of simply repairing the highway and leaving the area beneath it as a low income 'ghetto', a number of academics, civil engineers and planners formed the Cheonggyecheon Revival Research Group (CRRG) in 2001.[4] The CRRG saw the potential to 'restore' what they perceived as a 'decrepit public space' in order to 'create a waterfront in the downtown, improve the environment and restore the historical value', and revitalize the area in a manner that would 'enhance growth momentum and attract more private investment' (Seoul Solution, 2014). The CRRG succeeded in catching the attention of Seoul mayoral candidate Lee Myung-bak, who publicly backed the proposal during his election campaign in early 2002. After becoming elected, Lee made the revival of the area a priority and established a technocratic 'triangular governance system', comprising the Citizens' Committee for Cheonggyecheon Restoration, the Cheonggyecheon Restoration Project Headquarters and the Cheonggyecheon Restoration Research Corps (Hwang, 2005, p. 11).

One local response was the establishment of the Cheonggyecheon Merchants' Coalition, which sought to gain compensation and relocation for small-scale street vendors who would be displaced by the mooted development. The organization met with limited success and Yoon (2022, p. 15) has identified that the 'revival/restoration' project decimated the livelihoods and opportunities of the area's established population:

> The Cheonggyecheon flea market, along with the Hwanghak market, essentially vanished after the reconstruction, and only 700 street vendors were left in the Seoul folk flea market by 2011. In other words, reproduction of local economy and day-to-day life perished to be replaced by structures of private urban development.

The technocratic approach to the rehabilitation of the stream involved the restoration of the stream's flow by pumping in water from the Han River and its tributaries and extensive landscaping of the waterside area, including the construction of extended walking routes. These developments were undertaken with the expectation that the private sector would respond by developing new businesses around the stream. The project commenced in 2003 and was completed in 2005 at a cost of *c.*US$386 million (Kim, 2020) and rapidly attracted international attention, being awarded the Best Public Administration Award at Venice International Architecture Biennale in 2004 while construction was still in progress. But there was also considerable local unhappiness about the manner in which various elements of the historic built structure of the area were being dispensed with rather than preserved. The establishment of the Cheonggyecheon Museum, designed to commemorate the long history of the area was also marked by controversy and compromise over whose version of that history was being prioritized (see Whitehouse, 2024).

Despite criticism, the reconfigured, 11-km-long stream zone was launched with a major opening ceremony in February 2005. The continuingly highly positive international assessment of the project is typified by Harvard University Graduate School of Design's Urban Case Study summary (2025):

> The evolution of the Cheonggyecheon River in just twenty-nine months from an outmoded utilitarian highway into a multipurpose performative infrastructural piece of unprecedented size merits recognition as a seminal project in contemporary urban design. The project is a remarkable achievement that recovers the biological and social ecology of the city, and demonstrates the profound ability of design at the urban scale to provoke positive transformation effectively over large territories. The project signifies a broader sea change in Asian attitudes toward city design, from a quantitative model concerned primarily with growth to a more qualitative program that incorporates quality of life and environmental sustainability into economic development strategies.

Missing from this statement is any recognition of the effective erasure of the previous socio-cultural character of the area and of the gentrification that followed, which changed the area's affordability and demographic profile (Lim *et al.*, 2013).

Despite the types of concerns noted above, the positive ecological effects of the project have been widely acknowledged. One of the most balanced assessments was provided by Lee *et al.* (2020) who identified:

> Although the project did not fulfill the conditions of ecological restoation, there were a few ecological effects that this project brought. The reconstructed Cheonggye stream has enhanced species diversity and decreased the percentage of exotic species compared to an unrestored urban stream. Also, the species composition of the reintroduced riparian vegetation of the Cheonggye stream has since grown to more closely resemble the natural reference streams, whereas dissimilar to the degraded urban reference site. The re-established vegetation has attracted numerous animal species, and thereby increasing the size and diversity of the biota from 17 species before reconstruction to approximately 400 species. Consequently, the naturalness of the reconstructed Cheonggye stream, as indicated by morphological and ecological characteristics, has improved markedly. In addition, the reconstruction project also improved the water quality.

The caveats in the above statement are significant. They remind readers that (as per such projects in general) the rehabilitation process had not meaningfully *restored* pre-urban ecology rather than re-introduced a variety of species present in 'natural reference streams'. While the rehabilitated stretch of stream approximates aspects of its reference models, it remains an artificial, isolated, environmental fragment that is not connected to nearby areas in a manner that allows the species and spaces to be networked and interactive in any meaningful way (Fig. 4.2).

Fig. 4.2. A central stretch of the Cheonggyecheon in early evening, showing side lighting to encourage night-time use of the amenity. (Jiyoun, Wikimedia Commons) (2012).

Over the years since it opened, the area has become something of an 'oasis' for the residents of the inner-city during their leisure hours and a significant tourist feature for the city, attracting an estimated 16 million visitors a year (Chosunilbo, 2025). Its appeal substantially derives from the widely recognized calming effect of water flows and adjacent green spaces (Nichols, 2014; Tang *et al.*, 2024), and a mitigation of the urban heat island effect that is particularly noticeable during hot periods,[5] together with the provision of extended pathways along the streamside and its consequent attractiveness to walkers and to those seeking moments of quiet respite. As a result, the area is actively promoted as a key tourism location.

Whereas many of the small, resurfaced streams in Zurich flowing through parks and gardens resemble, at least to some extent, their pre-engineered, mid-1800s states through *not* having their banksides landscaped to accommodate paths, sitting areas, memorials, signage, etc. and by having naturalistic plantings; contemporary Cheonggyecheon bears little resemblance to any previous state of the stream and its surroundings. Kim (2020) has provided a nuanced analysis of the function and aesthetic of the rehabilitated area in terms of landscape urbanism highlighting the role of the imaginary in landscape urbanism:

> As a landscape and urban design project, the Cheonggyecheon illustrates time through both actual images, in photographic and other art installations, and through a green space design that serves as a counterpoint to the urban condition surrounding the space. Collectively, it is through an act of the imagination

> that together the sunken pedestrian channel of the park and the stream accomplish a transformative leap of form and time by breaking up the contemporary urban landscape, and its evocation of the new stream is at once a reflection of a time and place before the modernization of Seoul.

The singular 'act of imagination' involved might have been better expressed as a compound one: the imaginations of administrators, planners, architects, landscapers, builders, etc. and the imaginations of those who have frequented the stream and/or become aware of it through various representations. In this context, the resistance work *Drifting Producers* (2004) produced by members of the flyingCity collective, which was established in Seoul in 2001, is notable. Influenced by the ideas of Guy Debord (previously discussed in Chapter 1), the collective was formed in response to the rapid pace of urban renewal in the city and a concern to document and support the efforts of various communities, craftspeople and businesses to weather the transition being imposed upon them (Scholte, 2020, pp. 145–146). Along with promoting psychogeographical maps in local art workshops, the collective created an installation that combined suspended models, data sheets and artworks, and a screening of videos of street protests about the marginalization of craftspeople from the newly constituted streamside. The work contested straightforward notions of progress along the waterway. It is salutary that more recent artworks in the area have tended towards more apolitical expressions such as the Cheonggye Soul Ocean Media Art event that ran in the latter half of 2025 involving multiple source projection in and upon the surface of the stream at night, creating moving image works that covered the surface in large blossoms, brought digital sea creatures into the shallow fresh water stream and even – in an affectionate acceptance of metropolitan symbolism – introduced the city's pink tourism mascot, Haechi. The distance of these images and associations from the history of the stream and its bankside communities shows the very different civic function the stream now performs within the city.[6]

Like Kalamazoo, Seoul's metropolitan authority has begun to consider the impact of climate change on its central water feature. While it has been rarely highlighted as one of its key functions, the city factored in the need for Cheonggyecheon to serve as a floodway (intended to rapidly move peak waters through the capital without overspill) into its daylighting design specifications in the 1990s. But it planned on an assumption of major flood events happening every 200 years (Seoul Solution, 2021). Acknowledging that this may have been an optimistic scenario, water retention basins, drainage tunnels and pumping stations are now being constructed in the city (Yang-heok, 2025) and modifications to Cheonggyecheon are also likely to be necessary.

4.4 Diffusing Daylighting

The Cheonggyecheon project has often been cited as influential on subsequent international waterway revitalization proposals and projects,[7] although precise research on any direct influence has yet to be presented and it is likely that it was one of a number of projects – of differing types in differing locales – that have offered models for subsequent urban redevelopments. In the late 1990s, for instance, another urban rehabilitation project began to develop on the opposite side of the world, on the western shore of an island in the Hudson River that involved the reconceptualization and what has been termed 'parkification' of a section of an abandoned railway viaduct that ran close to the shoreline on the western side of Manhattan. The project was initially conceived and promoted by a small group of local activists keen to turn a derelict aerial eyesore into a local amenity and – like the rehabilitation of the Cheonggyecheon and its adjacent area – gained major traction when a mayor, in this case, New York mayor, Michael Bloomberg, began actively supporting the project in 2003. Now known as the High Line, the development, which comprises an elevated walking track, accompanying patches of vegetation and sitting areas, opened in sections between 2006 and 2014 (with a major new stretch, known as the Spur, opening in 2019). While the linear sky-park has been praised and embraced by New Yorkers and visitors alike, it has also been the cause of concern and critique due to its gentrifying effect on the areas it runs through, which have seen marked rises in property rental and purchase costs since its inception. Similar factors have arisen from a lower profile project that commenced in 1998 (and is still in-progress) paralleling the lower section of the High Line, the Hudson River Park.[8] Although the river park is a more conventional regeneration exercise than its elevated companion, the combination of the two have transformed a previously run-down area of the metropolis into a desirable location. It is hard to identify a directly causal relation between the High Line redevelopment and other rehabilitation projects in New York, and across the United States more generally, but elements of the Saw Mill River redevelopment in Yonkers, located 23 km north east of the High Line, invite comparison to the viaduct park and embody an enhanced type of the blue–green design to that previously developed in Cheonggyecheon.

The Saw Mill is a short river that flows down to Yonkers from a pond located 38 km to its north-east. In the late 1800s, it became increasingly polluted by industrial operations along its banks and was extensively culvertized in its lower section to conceal its noxious and malodorous flow. Calls for the daylighting of the lower portion of the river and the cleaning up of its polluted water commenced in the late 1980s and in the 1990s work was done to improve water quality and some paved-over areas of the river were opened up between 2010 and 2012 (Fig. 4.3). The most striking feature of the rehabilitation project involved the daylighting of an area to the immediate east of the suburb's railway station and the diversion of its culverted flow into a new channel with feature ponds and accompanying ladders and oxygenating cascades intended to allow regional fish populations to navigate it. This formed the central feature of a new landscaped area designated as Van der Donck Park. As Stevenson (2024, p. 276) emphasizes, this project 'never attempted to return the river to its precolonial state but to restore ecosystem services and

Fig. 4.3. Section of daylighted Saw Mill River near Getty Square, Yonkers. (Wikimedia Commons) (2023).

function within an urban environment', and was significantly successful in the latter regard.

Stevenson (2024) provides a critical case study of the motivations and planning of the Saw Miller River project, its physical elements and socio-economic outcomes and the manner in which they lacked the ethical element of place (re-)making and community care identified by Eckenwiler (2016). Stevenson's study focuses on what he terms the 'dark side of restoration' (2024, p. 265), which he attributes to the scheme being driven by local government, local business interests and investors with minimal acknowledgement of or consultation with Lenape (local first nation) groups or the African American and Latino communities that were prominent in the pre-development era, but have increasingly been marginalized by gentrification. He goes on to contend – in a phrase that could equally well apply to Cheonggyecheon – that 'the benefits, as real as they are, were never intended for the local population as much for new influx to replace them' (2024, p. 277). These concerns remain germane for similar projects elsewhere and point to a significant issue in both top-down civic projects and more mid-level ones where local groups and/or businesses initiate and nurture regeneration projects that are substantially taken up by local government. Yet, as projects such as Sao Paulo's Cidade Azul ('Blue City') have demonstrated, ambitious goals and enthusiasm are insufficient to deliver major urban redevelopments without a substantial buy-in from metropolitan authorities.

Launched in 2015, the Cidade Azul project developed out of Brazilian journalist Denis Russo Burgierman's fascination with tracing the paths of *ghost arroyos* (forgotten/buried rivers) across the contemporary cityscape. Unlike the writers, urban explorers and *ankyo* devotees whose work was covered in previous chapters, Burgierman wanted to draw on this knowledge to regenerate aspects of his city. With minuscule funding, the Cidade Azul project produced maps, posters, audio guides and videos about the *ghost arroyos* and identified its educational activities as 'an important step, towards change, towards understanding the city's dynamics' that could inform public policy and with the 'final dream' being stated as to 'bring these rivers to the surface' and provide water transport along them and gardens and cycle tracks along their banks (Montagner, 2015). While the project has continued to produce educational materials, its most ambitious aspirations have not come to fruition, given a lack of interest in committing the necessary resources to the project at a time of severe economic pressure in Sao Paulo and ongoing difficulties in supplying basic water needs to its inhabitants (Zambiasi, 2025). In such a context, it is possible to see how daylighting might be considered something of a luxury, however much it might benefit the city's inhabitants by providing attractive and healthy blue–green spaces. Similarly, a number of more formally argued proposals for daylighting projects that draw on the seminal international examples cited above (such as those advanced by Delibas and Tezer, 2017 for Istanbul's Ayamama Stream), have yet to achieve any traction. Reflecting this, only a handful of small-scale projects have come to realisation in the Global South over the last two decades (and these examples derive from its most developed nations, such as Taiwan[9]) and river daylighting thereby remains a predominantly post-industrial, Global North enterprise. It is also notable that while extensive riverside rehabilitation and 'renewal' projects have been undertaken in multiple locations, several of which resemble Cheonggyecheon,[10] no daylighting project of a similar scale has been implemented in any location since the completion of the Seoul project. Indeed, it might be argued that the Cheonggyecheon's most significant influence has been on blue–green riverside rehabilitation projects in general, rather than on the smaller scale, more locally focused projects that have proliferated in recent years.

Discussions with colleagues in the River Cities Network (an international organization with 44 member groups across five continents) in the lead-up to writing this volume proved illuminating in that there was limited interest in both the history and concept of daylighting and its potential for urban enhancement. This is, in part, explicable, given that many of the groups concerned were not aware of substantial culverting in their areas that might merit subtractive modification. It also reflects the limited traction that many groups have gained in pursuing less ambitious and lower-cost rehabilitation projects in their locales. But another factor, voiced by several discussants, concerned their perceptions of the impact of the Cheonggyecheon redevelopment on the area's established community. In contrast to those urban planning bodies that have asserted Seoul's redevelopment as an exemplar of the 'refreshment' of inner-city environments, the displacement of the area's population and the lack of meaningful local consultation involved in the process have been perceived as the project's most enduring legacy by some, providing a striking cautionary tale about the impact of large-scale, top-down developments on communities. It is notable, in this regard, that issues concerning river daylighting, associated blue–green rehabilitation exercises and the need for these to relate to urban histories, cultures and local neighbourhood concerns have increasingly been asserted by community organizations, researchers and urban authorities. The following chapter profiles a number of socially engaged ways of engaging with, imagining and reconfiguring social relationships with urban waterways that serve to bridge the gap between the types of affective engagements and representations profiled in Chapters 1–3 and the large-scale urban planning projects documented above.

Notes

[1] See, for example, Charbonneau and Resh (1992) and Charbonneau (2000).
[2] See Damodaran (2021) for discussion.
[3] Markets where second hand, cheap and/or artisan produced goods are sold at low costs.
[4] See Seoul Solution (2014) for further information on the early development of the group.
[5] There is some dispute as to whether tree shade along the river or the river itself is the most significant cause of the cooling effect, with microclimate modelling by Herdt *et al.* (2017) suggesting the former as the most significant factor.
[6] See https://www.facebook.com/korean4topik/posts/-exhibition-on-the-water-cheonggyecheon-media-art-2025-seouls-iconic-cheonggyech/1403294945130474/
[7] Sibilla, Sciandra, Rosso and Lamera's 2016 proposals for extensive daylighting of covered-over areas of the Navigli canal system that encircles central Milan, notably identify Cheonggyecheon as 'a guideline for the daylighting of urban man-made canals' (2016, p. 248).
[8] For a historical overview of the development, see Hudson River Park (2020).
[9] See Wantzen, Piednoir, Cao *et al.* (2022), pp. 19–20.
[10] See, for instance, those profiled in Hayward (Ed.) (2024).

5 Hydrocitizenship, Heritage and Cultural Daylighting

5.1 Introduction

We began this volume by discussing sociohydrology as a relatively new term and concept within hydrological discourse. In this section of the final chapter, we explore aspects of another relatively new appellation, *hydrocitizenship*, that characterizes social care and concern for waterways and water issues more generally. While hydrocitizenship can be considered as a sociohydrological enterprise, the latter should not be understood to subsume the former but, rather, to offer a context for it. Social engagement with issues concerning water quality, water management and the general role of water within inhabited environments has been evident at various stages in human history and particularly so in response to rapid transformations of hydrology within industrialized societies and landscapes in the modern era. Hydrocitizenship can be understood to encompass social roles and responsibilities with regard to water in all its forms. The term appears to have emerged in the late 2000s and has been increasingly used in various British contexts such as the UK Arts and Humanities Research Council funded, multi-site 'Towards Hydrocitizenship' project[1] that ran in 2014–2017. Graeme Evans, a chief investigator on the Lee Valley component of the project has explained the core concept in the following terms:

> Within social-hydro relations, as in other contexts, citizenship can be understood in diverse, contradictory ways, including, for example, as a social contract, and an individual or moral responsibility which impacts on the practices of a community. There is a need to investigate and understand how particular socio-hydrological configurations... are produced and how these contribute to potentially inequitable environmental conditions. Water issues generate specific and contentious forms of conflict within the eco-social fabric, which demand exposure and negotiation in any form of culture, and necessitate policy for addressing rights, sustainability and resilience. Hydrocitizenship is thus conceived as an enhanced awareness of, and sense of responsibility for, water as a vital, shared ecosocial resource. It implies ongoing processes of learning, negotiation and practice rather than a precisely defined/static position... It also seeks to engage

> directly with critical issues of global sustainability and environmental change in local/community settings by envisioning these 'macro' issues through the lenses of everyday relations and lived experience. (2018, p. 203)[2]

The concept of hydrocitizenship is one important way of understanding various river-focused initiatives that have risen in prominence since the 1960s and for approaching and evaluating proposals for and outcomes of daylighting projects. This can also be intertwined with King's concept of the 'hidden hydrology' of subterraneanized flows (2025), and his identification that engaging with these 'offers new ways to conceptualize a range of interventions' drawing on different perspectives produced by art, landscape architecture, urban ecology and planning to produce deep social engagements with rivers and water resources more generally. As Evans (2018, p. 209) also identifies, this involves contestations over and dialogues about how various flows and spaces can be conceived and negotiations over the manner in which stakeholders should be able to influence modifications to hydrological systems and related water/land-scapes.

Hydrocitizenship reflects a close affinity felt by (some) individuals and communities with regard to water and waterways (including buried ones) that has been characterized as hydrophilic. While the term 'hydrophilia' has multiple meanings (including various bio-chemical attributes and processes), it has increasingly been used to refer to deep-seated human affinities with and desires to be in the proximity of water in various forms (rivers, canals, ponds, lakes, fountains, coasts, open seas, etc.).[3] This is manifest in various aspects such as leisure, well-being and various forms of environmental inspiration and perceived tranquillity (aspects highlighted in Foley *et al.*, 2019). Urban exploration and related research (of the types profiled in Chapters 1–3) and social visitation of daylighted spaces (as discussed in Chapter 4) can be understood as 'hydrophilic encounters' (Kitson *et al.*, 2023, p. 780). The authors describe these as 'moments of perceptible awareness of feeling connected through water' (2023, p. 780). Reflecting on their own experiences with running public storytelling workshops in New Jersey in 2022, they assert that experiential, practical and participatory 'encounters' with water are crucial for enhancing a sense of place within a community, as they 'reveal complex fluvial senses of place situated in the past, present and future' (2023, p. 780). While often overlooked in policy and planning discourse, these felt experiences and sensibilities are core to many of the initiatives discussed in this chapter.

Whatever conceptual frames are applied, individual and collective care for water systems has developed apace over the last two decades. Aside from the pioneering urban projects discussed in Chapter 4, research for this volume suggests that a significant inspiration for locally attuned river daylighting schemes in the early 21st century has been the promotion of the concept through print, broadcast and online media. One notable trailblazer was a feature article by Andrew Revkin in *The New York Times* published in 2009 headlined 'Peeling back pavement to expose watery havens'. The article focused on the Cheonggyecheon development as a seminal moment in urban daylighting and outlined the benefits of the practice. Another notable media contribution was Caroline Bacle's feature documentary *Lost Rivers* (2012). Since its release, the film has been widely praised, selected for a number of festivals and frequently screened with accompanying presentations and discussions by river researchers and urban planners.[4] A number of journalists have also drawn on Bacle's film, such as Cox (2017), in *The Guardian*, in an article that proclaims daylighting as a 'global movement'.

Similar to our focus in this volume, Bacle's documentary combined sequences shot inside buried rivers, sequences of urban explorers inside them and sections on seminal river daylighting projects. Bacle has described the initial inspiration for her film as her encounter with the activities of Montreal-based urban explorer Danielle Plamondon. As previously outlined in Chapter 1, urban explorers (urbex) secure access to prohibited buildings and subterranean spaces and often document their explorations on video. Bacle's encounter with Plamondon's work alerted her to the rivers buried beneath Montreal's modern cityscape, prompting her to enter and film Riviere Saint-Pierre. Plamondon also acquainted her with the activities of urbex groups such as Bresciaunderground APS in the Italian city of Brescia, who have led visitor tours of ancient hydrological structures, who are also documented in the film. Along with visually striking representations of underground tunnels and of the constrained nature of (once) free-flowing rivers, the film presents a largely positive representation of daylighting projects such as Seoul's Cheonggyecheon and Yonkers' Saw Mill River, showing the processes and results of subtractive engineering and riverside redesign. The documentary covers a wide range of material in a relatively short duration, but appears to have been effective in instilling enthusiasm for the revitalization and re-presentation of buried rivers, inspiring local activists to explore and propose similar schemes, particularly in the United Kingdom (see Section 5.3 below).

The following sections offer accounts of hydrosocial projects that created communities of interest among activists and local residents whose discussions and consciousness-raising were significant in their own right; creative approaches to stimulating such discussions; and accounts of successful heritage-oriented daylighting projects that resulted from active team and coalition building.

5.2 Local Activists and Community Consciousness

As detailed in Chapter 4, various river daylighting projects have been successfully implemented over the last 40 years. There have also been a series of less visible and, thereby, largely overlooked initiatives that either never coalesced into solid proposals or else have not been supported by relevant authorities. Similarly, there have been other hydrocitizenship projects that have – largely unsuccessfully – aimed to resist the ongoing culverting of surface flows in particular locations. Despite the failure of many of these to deliver significant structural outcomes, several of the projects that have been documented and analysed by participant researchers have been insightful for broader approaches to hydrocitizenship.

In a study evocatively titled 'The Time of Ghost Rivers', Kregg Hetherington (2025) offers an account of his involvement in a project to prevent a single remnant surface stretch of Montreal's Meadowbank Creek from being channelled into a pipe in the early 2020s. The account is valuable as a participant-observers' reflection on the project and the assumptions and experiences informing it. After characterizing the Friends of Meadowbank as a highly motivated, educated and affluent group capable of gaining support and exposure from various professionals, Hetherington offers an insightful characterization that is just as apposite for chroniclers of subterranean ghost rivers as those intent on preventing their burial.

> What they were doing was infrastructural inversion, the hermeneutic flip whereby the network supporting a thing becomes suddenly visible, sometimes more visible than the thing itself. Since it was never going to be possible to simply save the creek without attending to where it came from, their research traced the many ecological, infrastructural, legal, and historical relations that made the creek what it was. But in the process of the inversion, the creek became something more than it had been, coalescing into a new kind of entity, a more complex presence, a life lived full of stories and attachments. The creek, a picturesque feature of a beloved golf course, became a river, an entity older than settlement itself. Rivers, particularly in the colonial imaginary, are semipermanent entities that can be mapped, like mountains and borders, onto static, twodimensional representations of nature. This river brought an added temporal twist: It had only become visible because it was mostly destroyed—a ghost river, conjured from the purgatory between life and death in anticipation of some point of no return. (2015, p. 672)

Following a protracted legal case that centred on a key semantic issue of whether Meadowbank Creek should be considered as a river or a ditch (with the judgement eventually selecting the latter category – Hetherington, 2015, p. 660), the appeal against the culverting was lost. Somewhat unusually, this did not lead to a sudden dispersion and deflation of the group but, rather, their planning and staging of a funeral for the Creek. As Hetherington has contended:

> One can attempt conjurings, exorcisms, reanimations, or exhumations, but in the incompletion of all of these, one is at least participating in what Derrida called giving ghosts a "hospitable memory," an acknowledgment of death "out of concern for justice." (2015, pp. 672–673)

Albeit addressed to a significantly different situation, McEwen *et al.* (2020) offer related points to those offered by Hetherington with their description of the rationale, methodology, results and reflections on a project conducted by a multi-disciplinary team in Bristol (UK) in the mid-2010s investigating interest in the daylighting of tributaries flowing into the River Avon's New Cut diversion.[5] Inspired by the work of creative activists such as Lillian Ball whose 'eco-diplomatic' work in areas such as New York's South Bronx has involved empowering marginalized communities to exercise hydrocitizenship, and 'recognising people's complex relationships and holistic interactions with water' (Ingram, 2014), the team adopted a participatory action/practice-as-research method that included:

> interpretations of water as a source of health, comfort, and identity, as well as life, livelihood, and risk. Throughout, we aspired to connect and empower people, allowing our research to flow along different braids opened by individuals' participation with their own interests, practices, processes, and backgrounds. (McEwen *et al.*, 2020, p. 782)

This involved an unusually protracted process of 'community enabling, participatory mapping and participant observation in an emergent process' concerned with aspects of individual and social memory undertaken over a two-and-half-year period (2020, p. 783). The addition of Luci Gorell Barnes, a locally based participatory resident artist, to the team, resulted in what they perceived as the team's ability to enable

> participants to evoke senses of citizenship in their neighbourhoods... exploring people's relationships with place through creative, participatory, narrative/deep mapping sessions... using handmade and digital mapmaking evoked participants' stories, oral histories, memories, ideas, and perspectives about their local neighbourhoods and waterways, and water (dis)connections. (2020, p. 784)

Appropriately enough for such a project, the team have used liquid allusions to characterize:

> our participatory daylighting process as a "cascade," working with the metaphor of physical river "braiding." This is conceived as a series of "upwellings," "openings," and "re-openings": emerging, connecting, enacting, imagining, and reflecting" (2020, p. 785)

Similar to the approaches utilized by various writers, urban explorers and recreational enthusiasts to explore ghost rivers in London, Sydney and Tokyo discussed in Chapters 1–3, the team used contemporary and historic maps as 'pivotal' elements in participant conversations about river history 'as starting points for rehearsing memories, in helping to make geographical and other connections over time and place, and as receptacles for stories told'. (2020, p. 787). One aspect of the project that the researchers identified as particularly productive concerned the familiarization of school children with previous daylighting schemes through audiovisual representations, leading them to produce a range of designs for their local waterways that were then discussed with the team and presented to local residents at dedicated meetings. The ultimate success of the project was not in any firm set of proposals regarding specific daylighting projects in the area but rather through the consultative and imaginative processes involved in considering such modifications to urban hydrologies in which 'daylighting' is as much a discursive and empowering exercise as any re-engineering of buried flows.

5.3 Heritage-based Daylighting Projects

One example of an initiative directly inspired by Bacle's previously discussed film *Lost Rivers* is Sheffield's Sheaf and Porters River Trust (S&PRT) established in 2019. Ric Bingham, an engineer and activist with S&PRT has identified *Lost Rivers* as the Trust's 'inspiration' (pers. comm, 2025). Simon Ogden, S&PRT's founding chairperson and (then) Director of Regeneration for Sheffield Council, had the film screened at Sheffield's Showroom Cinema for the Trust's inauguration in 2019. As he has also identified

> Film played a further role as we initiated our Hidden Rivers Tours first as a part of the annual Sheffield Adventure Film Festival. The earliest tours even included a short film show in one of the drier culvert tunnels. More importantly it established our relationship with a team of professional caving guides who allowed us to overcome official resistance to accessing the culverts with public tours. (pers. comm, 2026)

Fittingly, with regard to its influence on local campaigners, the S&PRT project has combined various of the activities featured in Bacle's film. The Trust's central aim has been to restore the presence and social awareness of Sheffield's buried rivers and to create new waterside environments to complement this. The tours referred to above have been a key aspect of the Trust's operation. As Ogden recounts:

> We have taken over 11,000 people on these tours, establishing it as a well-known 'bucket list' experience and generating thousands of pounds which have enabled our other less glamorous activities such as clean-ups, training citizen scientists and producing visioning documents... The tours also provide a useful platform for engaging politicians, young people, riparian businesses, developers and public bodies including Fire and Rescue, the Environment Agency, Yorkshire Water, Transport Police and the Planning Inspectorate. (pers. comm, 2026)

Its awareness of the logistical issues involved in daylighting urban rivers has also led it to seek novel and pragmatic compromises. With specific regard to the area of the Sheaf that was covered over to allow the construction of Sheffield Railway Station in 1868, for instance, the S&PRT has identified that

> whilst removal of the culverts under the railway tracks of Midland Station is probably unrealistic the potential of sinking light-pipes into the dark tunnels to allow fish and animal passage could be considered as part of the environmental mitigation of the scheme. (S&PRT, 2019)

S&PRT's activities have been multiple and have included planning and campaigning for the re-opening of stretches of the Sheaf and Porters Creek and bankside redesign; negotiating with the station management about the previously mooted installation of light wells; organizing guided tours of the

Fig. 5.1. Exploring the Megatron Tunnel. (Location Unknown UK, 2022). Figure used with permission.

Megatron and Sheaf Culverts under the city centre;[6] and providing an online time-lapse video of the deculverting of a substantial slab over the River Sheaf near the city's Exchange Street.[7] The S&PRT has implicitly recognized the buried Sheaf and Porter Creek as hydrosocial territories split between terrestrial and subterranean levels and has advocated their rehabilitation as beneficial to regenerating key areas of the city and creating a biologically active blue–green corridor through its centre (see Fig. 5.2).

Another notable local-heritage-oriented project concerned the River Roch in central Rochdale (in Lancashire, UK). Rochdale developed as a major centre for woollen textile production in the late 1700s when factories replaced dispersed cottage production networks across the region. It was closely integrated with Manchester and with the Calder and Hebble Navigation waterway in Yorkshire by the Rochdale Canal, which opened in stages in the early 1800s. As the century progressed, cotton fabric production grew in importance and the amount of trade and traffic within the town increased, resulting in the construction of a new bridge across the river in 1882, relatively close the town's older, modified medieval bridge. The river, which once offered a navigable route through the town was increasingly neglected from the mid-1800s on and was essentially treated as a sewer. In the early 1900s various sections

Fig. 5.2. Visualization of proposed river exposure and rehabilitation of former Castle Market site in Sheffield showing 'the right bank of the river with much of its concrete wall re-profiled to provide a series of terraces down to almost water level... The river channel is also transformed by the removal of the 2-m-high Castle Orchard Weir to create a series of gently rising pools, which will allow fish and wildlife to traverse this reach of the Sheaf to and from the Don, as well as creating opportunities to revegetate the formerly sterile concrete banks and providing habitat for birds, insects, fish and mammals'. (S&PRT, 2025). Figure used with permission.

between the two bridges were paved over, finally covering an area sufficient to establish a new tram stop in 1924. While this disappearance of the river from the town centre appears to have been little lamented at the time, there was a revival of interest in the concealed bridges and the river that ran underneath them following the removal of surface coverage during culvert repair works in the mid 1990s that allowed council workers, reporters and photographers the opportunity to glimpse and represent the architecture of cavernous spaces beneath the town's streets before they were closed over again.[8] An initial campaign to have sections of the river permanently daylighted failed to gain traction in the late 1990s but the issue persisted over the next 15 years, as similar projects began to occur in other European and North American locations. In 2010, Rochdale Council commissioned an architectural report that identified the medieval bridge and more modern concrete culvert as a significant historical artefact. This assessment, together with broader council commitments to manage surface water flows and flood risk and to revive the town centre, led to the design and proposal of a project titled 'Revealing the Roch'. This proved successful in attracting public funds, securing *c.*5 million UK Sterling from the Heritage Lottery Fund, Environmental Agency and the North West Regional Flood and Coastal Committee in 2015. The funding allowed sections of the river and the side of the modified medieval bridge to be exposed, restored and encircled by pedestrian areas, seats and tree plantings, forming a public space that opened in 2016 (Fig. 5.3). Since the daylighting project was undertaken, surface water has begun to drain into the exposed river and flood events in the centre have been substantially reduced.

One notable aspect of the Roch project is that it included promotional activities in its early stages that tried to build local support for the project by disseminating information about the area's history and the heritage value of the medieval bridge and the restoration process. Indeed, Connelly *et al.* (2019, p. 143) have gone further by contending that

> The physical works provided an intuitive link to the pre-industrial buildings located in the town center, which had long since lost their orientation towards the hidden river.

More broadly, the authors have contended that the overall project proved highly successful for the town and its council's sustainability orientation by 'uniting economic growth with environmental and social objectives as the so-called "triple bottom" line;' illustrating how culture might be considered as 'sustainability's fourth pillar' (2019, p. 141). Picking up these themes, we move on to related concepts of 'cultural daylighting'.

5.4 Cultural Daylighting

Along with the projects outlined in Section 5.2, another significant engagement with local residents was carried out in Manchester in 2023 in areas adjacent to

Fig. 5.3. Daylighted area of the River Roch adjacent to Rochdale Bridge. (Matthew Wells, 2025). Figure used with permission.

sections of the River Irk culverted in the Victorian Era. The project was led by environmental journalist and researcher Joe Shute and, unlike the previously discussed Avon New Cut project – which Shute refers to in his discussion – was focused on what he terms 'cultural daylighting', an approach intended to 'resurface' *stories* of rivers and enable residents adjacent to engage with them in a manner that allowed them to 'unearth and develop stories attached to rivers and in the process reconnect people to lost waterways' (Shute, 2024, p. 609). Indeed, Shute contends that 'community engagement with place holds particular value against a backdrop of urban regeneration' that homogenizes rehabilitated waterways in ways that facilitate displacement and gentrification (2024, p. 609). In this context, he references psychogeographer-folklorist John Billingsley's advocacy of 'a subtle understanding of place, assessing esoteric as well as exoteric associations, undermining the alienation of modern-day planned landscapes' (Billingsley, 2017, p. 17). Shute's project involved the establishment of seven collaboratively designed workshops with different generational compositions that attempted to promote a sense of communal history. Working in a low socio-economic status area with low levels of educational attainment and with mixed levels of English Language proficiency, he avoided prioritizing written inputs and outcomes in favour of using storytelling, textile work and clay-modelling 'to empower and embolden participants to explore their own relationships to the river'. (2017, 619). One of the issues, Shute had to negotiate was that given

> the current level of disconnect between the river and the communities it serves, many of the younger participants I have been working with do not have their own stories of the river, and indeed I discovered in many cases have never even previously heard of the Irk, or knew of its existence. (2017, pp. 617–618)

In order to overcome this knowledge gap Shute used what he terms 'story-swapping' between different groups 'using anecdotal or archival histories' such as fishing in the Irk 'to pass folk memories of the Irk along the generations and inform our work' (2017, pp. 617–618).

Shute perceived that one of the key takeaways from the project was that a lack of common knowledge about a neglected and/or substantially buried river such as the Irk

> does not restrict a neglected river from becoming a cherished community asset and focal point of a diverse range of intergenerational groups. During the workshops, even those who had never heard of or visited the Irk were able to draw on experiences of other rivers, including from as far afield as Indonesia and Kashmir, as inspiration for their creative work. The lack of a sense of shared heritage does not necessarily pose a barrier to developing shared visions of future change... Regardless of a lack of shared history around the River Irk, the participants involved in the workshops developed clear links in terms of what they would like their future engagement with the river to be. The process of culturally daylighting the Irk and eliciting stories from the past ensures that the river can once more be a focal point of the communities it flows through; a reclaimed natural asset re-established in the folk memory of Manchester. (2017, p. 618)

Shute's flexible workshop activities aimed to stimulate participants to engage with the 'lost' River Irk by drawing on multiple perspectives within workshops. Another, alternative and more provocative approach to stimulating local engagement with submerged riverine histories and alternative urban futures was pursued elsewhere in England, in Coventry, by postgraduate students from Sheffield University's (Re)Activist Studio in 2017. The students' external intervention was conceived as a form of 'culture bombing' – a technique of disrupting conventional modes of perception and behaviour akin to Situationist *détournements* (cultural performances or texts that modify existing works or structures in an arresting manner so as to open up alternative readings or critiques).

The (Re)Activist Studio students' performance activities centred on raising awareness of the River Sherbourne that ran through the Coventry until the 1960s when it was extensively culverted to facilitate road and building construction over its former route as part of a major urban redevelopment project. Unlike the case of the River Irk, subterraneanized in the late 1800s, the burial of the Sherbourne occurred within the lifetimes of older Coventrians and thereby lingered more strongly than the submerged histories Shute sought to resurface around Manchester. Indeed, there had been active interest in daylighting the river since the 1990s, various incursions by local urban explorers and various projections of the financial benefits of doing so in terms of flood prevention advantages prior to the students' interventions (Chambers, 2015).

The students used what its designers referred to as a 'Kino-Cine-Bomber' apparatus in acknowledgement of Soviet radical film maker Dziga Vertov's dynamic concept of 'kino-eye' film-making, which sought to de-familiarize and juxtapose representations of everyday life in ways that provided insights into modern experience.[9] The principal difference between the (Re)Activist Studio approach and its stated inspiration was that the former relied on a mobile, bicycle-based, image projection system that the students pedalled around the city and used to project their alternative visions of urban space use around the buried river onto buildings and walls. These visions were predominantly subtractive, in the sense developed by Easterling (2003), involving the deconstruction of buildings and built surfaces in areas such as culverted rivers, riversides, floodplains, etc. so as to reconfigure built spaces and restore points of integration with prior landscapes, physical features and ecologies. More specifically, the project's coordinators aimed to use students' initial research data about the city and projection events

> to help identify infrastructure and buildings no longer fit for purpose that could be subtracted from the urban landscape, where segments of the River Sherbourne could be uncapped, revealed, and celebrated as urban leisure features, contributing to the social, cultural, and economic vitality of the city. (Lashua and Baker, 2019, p. 699)

The Kino-Cine-Bomber approach merits further exploration as a complement to more conventional approaches to imagining, proposing and acquitting daylighting projects and informing discussion as to whether these may be the best options for neighbourhood or more extensive urban renewal and environmental enhancement.

5.5 Ghostly Territoriality and Riverine Personhood

One of the most imaginative and insightful characterizations of the complexity of modern riverine hydrosocial territories that include culverted components has been provided by Cotoi, Iorga, Păduraru and Georgescu (2025). Their study of Bucharest's Dâmbovița River not only provides an account of a complex, dual-tiered flow that includes an upper river section and lower sewerage channel but also develops this within an understanding of the contemporary Dâmbovița as complex space that

> is not merely plural, as in different beings living side by side inside a homogenous time and space, but multiple, existing through complicated relations of interference, inclusion, representation and historical sequence. (2025, p. 1448)

Within this, they argue, its subterranean section 'constitutes a form of hidden urban leakiness – one that complicates the historical–political binaries of socialist ghosts vs postsocialist capitalist nature and creativity' (2025, p. 1449).

As the authors identify, the 'discursive space surrounding the Dâmbovița is populated with images of innovation, green–blue infrastructures and cool, breezy spaces... hovering above a seemingly lifeless watercourse'.

> More than just a huge administrative challenge, the Dâmbovița emerges as a multiple entity. The river is subject to various institutional, legal and moral interpretations and judgments, each attempting to map it, to place it within one or another urban and infrastructural frameworks. Competing cultural, technical, historical and political imaginaries shape debates over its role as a central urban hydrological infrastructure. At the core of these contestations lies a deep fracture, one that goes across all of the river's 'ontologies' ... its bareness and productiveness, its being both an enduring infrastructural ghost and a new—yet unachieved—environmental river. (2025, p. 1449)

As these discussions suggest, the dual-tiered urban river is, arguably, now such a highly engineered hydrosocial system that its pre-industrial, pre-metropolitan 'personality' – which persists in its upstream sections as it flows down from this source in the Curmătura Oticului mountain pass – is effectively irrecoverable. As the authors characterize, the river is 'part of the complicated historical-semantic rift and dynamic between the ghost and the living rivers' and 'amidst the present-day powerful environmental-oriented imaginaries and the active historical (mis)shaping of the past there is an audible murmur of the 'unaccounted for, unnamed, and whose fictions are only registered as noise" (2025, p. 1460).

The poetically interpretative nature of the authors' concluding comments address the suppressed identity of the original natural/wild stream that flows from the mountains and suggest the manner in which rivers have increasingly been perceived to possess personalities that – necessarily – derive from what have been characterized as their personhood. This returns us to a notable paper referenced in the Introduction to this volume, Lytheberg and Hikuroa's (2020) study of the culvertized Wai-Horotiu River in Auckland (New Zealand) (Fig. 5.4) and the ontological question of how we can *know* such flows, particularly in their buried and constrained forms.

Drawing on traditional Māori culture, Lytheberg and Hikuroa (2020, pp. 373–374) summarize their object of study as follows:

> Wai-Horotiu stream used to run down the valley that is now Queen Street, the main road in Aotearoa-New Zealand's largest city... Treasured by Māori as a source of wai (water) and mahinga kai (food), it was also the home of Horotiu, a taniwha or ancestral guardian—a literal "freshwater body." However, as Tāmaki-Makaurau transitioned into Auckland City, Wai-Horotiu became denigrated; it was used as an open sewer by early settlers before being buried alive in the colonial process. The containment of Wai-Horotiu within a barrel-vaulted brick sewer was simultaneously an act of progress for settler-colonial public health and the live burial of a living being, and therefore deep tragedy.

As this description immediately suggests, the buried river is understood as a 'treasured' 'live being' (in itself) and also as the 'home' of a *taniwha*, a

Fig. 5.4. Culvertized section of Wai-Horotiu River exposed by construction work. (Ingolfson, Wikimedia Commons, 2007).

powerful ancestral guardian. Its watery space is charged with being and is easily understood as having a personality. The authors ask how we can know this entity and what rights it has and provide one answer to their question in terms of the 2017 Te Awa Tupua Act that afforded the Whanganui River, in the south-east of North Island, 'juristic personality and moral considerability' (Lytheberg and Hikuroa, 2020, p. 374). The Act has been seen as a landmark in global environmental law for its legal recognition of the Whanganui River as a living entity with rights and received the World Future Council's 2025 Global Impact Award on that account. Extending a similar recognition to Wai-Horotiu (and to Aotearoa's waterways in general) provides a framework for understanding the stream's repression. Indeed, together with concerns about urban drainage following high rain events in the city in 2022 and 2023, it provides further impetus to calls for Auckland Council to revisit their (as yet unrealized) concept design for stream daylighting for central Auckland (2005). While proposals for the daylighting of stretches of the Wai-Horotiu feature in a report that somewhat problematically cites Cheonggyecheon as a 'best practice example' (Lewis, 2008, p. 6) and has minimal reference to Māori intangible cultural heritage and beliefs, it notes that its proposal for daylighting the Queens Street section of the Wai-Horotiu has 'the potential to reawaken the consciousness of the public to water in the cityscape, and to the "sense of place" that natural heritage can invoke' (Lewis, 2008, p. 6).

It would be interesting and, no doubt, highly disruptive to existing models of urban management, if the Dâmbovița, or other buried rivers discussed in this volume, could be afforded the same 'juristic personality and moral considerability' as the Whanganui River. The costs of any extrication that could be required as a result of successful legal actions conducted on their part might well stress the financial capacities of the cities involved. It is possible, in this regard, that arguments might be made to protect existing stretches of buried rivers from any litigation that aimed to deculvert them, but it would constitute a considerable advance if the rights of existing un-culverted rivers would have to be addressed in any subsequent proposals to culvert or otherwise divide or divert them. Recognition of rivers' rights would also provide a different ontological approach to understanding the essences of rivers – 'wild' or culverted – that might be reflected in a respectful and considerate consideration of their integrity, rather than the 'mere appendages to waterworks and sewerage systems' they were considered as in the early phases of metropolitan development (Castonguay and Evenden, 2012, p. 3) or as the unpleasant and inconvenient flows that merited burying away from public gaze or olfactory perception.

Notes

[1] The sites being in Bristol, mid-Wales, The Lee Valley and Bradford.

[2] NB while Evans and other project coordinators quoted in this chapter appear to refer to implicitly preconstituted communities as the subject of their interventions, we have reservations about such characterizations (in line with those offered by Brint, 2001 and Fordahl, 2022) and avoid them as far as possible, preferring to regard daylighting projects as community *building* initiatives.

[3] It should be noted that the opposite tendency, (social) hydrophobia, has been less-well researched and theorized but is also evident in various communities at various historical junctures and continues to inform aspects of spatial management and health and safety management (see Pitt, 2019, for example). Similarly, there is little discussion of the middle ground of 'hydro-indifference'.

[4] Indeed, the film has had considerable longevity in this regard, still being featured in 'Science on the Screen' events in the USA in 2025.

[5] An engineered waterway originally excavated to divert the main flow of the Avon in the early 19th century.

[6] See https://www.sheafportertrust.org/guided-tours

[7] See https://www.sheafportertrust.org/culvertcam

[8] See Flynn and Brooks (2020) for a description and photo documentation of these.

[9] This concept informed both his Kino-Pravda series of short Kino-Pravda newsreels made between 1922 and 1925 and his most acclaimed work, the 68-min-long *Man with a Movie Camera* (1929).

Conclusion: Urban Riverscapes and Decelerating Development

Ten years ago, a group of environmental scholars edited a collection of essays entitled *Arts of living on a damaged planet: Ghosts and monsters of the Anthropocene* (Gan *et al.*, 2017). Equally colourfully, they subtitled their introduction 'Haunted landscapes of the Anthropocene' (1–15) and opened it with the following characterization, '[t]he *winds* of the Anthropocene carry ghosts – the vestiges and signs of past ways of life still charged in the present' (1). While their anthology did not include considerations of rivers, the themes of damage, ghosts, hauntings, vestiges and remnant charges are clearly apposite for (and manifest throughout) this volume. Indeed, we might substitute their reference to the winds of the Anthropocene with the *watery flows* that threaten to disrupt the stable surfaces of metropolises built without premonition of rising sea levels or intensifying weather events. These swirling associations are not simply fanciful; they are the signature elements of the Anthropocene that frame our discussions. This situation is innately unsettling for (orthodox) hydrology, a discipline premised on the management of water within supposedly predictable parameters. While its scientific basis still pertains, the scale and dynamic range of the phenomena it is now required to address are challenging. Similarly, sociohydrology is also in transition as it moves from being a social-scientific adjunct to hydrology to a field that has to acknowledge and encompass such unquantifiable components as cultural hauntings and the return of repressed flows and energies.

This volume has attempted to contribute to the broadening of sociohydrology by discussing the rationales and processes of burying rivers; the resultant cultural comprehensions of such buried flows; and various impulses and strategies for their de-culverting and re-exposure. Chapters 1–3 focussed on buried rivers and waterways. The Fleet's relegation to a metropolitan underworld and its post-memorialization in various media accounts typifies the manner in which modern cities have 'overwritten' the geo-spatial features that once enabled and characterized them. In the work of writers such as Ian Sinclair, the Fleet is perceived to flow unseen beneath the streets of the city as a tangible – if largely unreachable – trace of the city's past that acts as a 'creative space in which feelings of loss and redemption are explored and negotiated' (Bonnett, 2009, p. 54). Sydney's Tank Stream, which experienced a similar burial has received far less attention, with its original Indigenous custodians so severely depleted and marginalized from metropolitan society and governance that they remain largely disconnected from its subterranean present. The case of the burial of waterways in Tokyo during various phases of 20th century modernization – and of the key individuals and media outlets involved in perpetuating fascination with them – is distinctly different from the undercurrent of public interest in London (and/or thinner trickle of attention in Sydney) and reflects a particular approach to metropolitan identity and social memory in the Japanese capital. These case studies outline the condition of buried rivers that have been the subject of attention without generating significant campaigns for their daylighting. Various factors appear to underly this. One is a tacit recognition that the costs and logistics of such daylightings would be prohibitive. But another, perhaps more significant, one is that the disconnection of the cities' ghost rivers from prior riverscapes is now taken as given.

Chapter 4 switched focus by discussing early urban daylighting projects that reopened stretches of Korean and American rivers and constructed linear public spaces around them. None of the main examples discussed either attempted to or incidentally succeeded in re-establishing their areas' pre-modern riverscapes. Similarly, they were not informed by sustained and insightful dialogue with local communities about local histories and the environments they wished to have access to. For all their successes in terms of public patronage, the top-down projects primarily constituted exercises in generic blue-green rehabilitation.[1] Chapter 5 provided examples of more consultative, creative and community-informed engagements with urban daylighting that can be understood as exercises in 'hydrocitizenship.' These focussed on socio-cultural – rather than merely functional-mechanical – processes. The chapter concluded by expanding scope and considering the implications of recognizing the 'juristic personality' and (quasi-) personhood of rivers on the perception and management of buried flows and the assessment of their potential for resurrection.

There are various ways of pulling together the threads of these discussions. The first is to reiterate that there is no necessary correlation between the intensity of socio-cultural interest in subterraneanized rivers and the likelihood of them being deculverted. The most 'ghostly' of ghost rivers – in terms of the extent/intensity of lingering perceptions and representations of their pre-burial phases – have a degree of stability and integrity as ghost rivers that is not obviously enhanced by their reopening and/or re-presentation. Nevertheless, the existence of buried rivers provides an opportunity for metropolitan authorities to undertake works that daylight their flows and to create engaging blue-green linear spaces around them. In a numerically small but significant number of cases, daylighting projects have been closely aligned to local heritage in manners that provide spatial access to and perspective points on buried waterways, and underground structures that arguably enhance their heritage resonance and create new and attractive surface spaces surrounding accessible deculverted areas.

As should be evident from the discussions advanced in this volume, buried rivers are complex hydrosocial territories that vary considerably from each other in form, function and significance for the populations of the areas they run under. There is, therefore, no simple, objective procedure through which a buried river might be adjudged as 'worth' daylighting and no consensus as to how 'worth' might be conceived in this context. That is not to say that particular daylighting projects have not been adjudged as successful by their proponents or by external agencies and indeed some – such as, most prominently Cheonggyecheon – have become major tourism features. Indeed, it might be argued that Cheonggyecheon now has heritage status as an iconic filament in the fabric of Seoul, within a global daylighting movement and as a leading example of large-scale, blue-green rehabilitation of an urban waterway. If we understand daylighting as an activity that can encompass a range of applications, Cheonggyecheon sits at one end of a spectrum with smaller scale, more locally attuned and heritage-sensitive projects such as those undertaken in Rochdale and Sheffield at the other.

Shifting focus to a planetary perspective, the timeline of river burials and attempts to daylight such flows clearly aligns to the emergence of the Industrial Revolution, the growth of major urban and metropolitan centres and the disruptions to all facets of the environment and the species within them. Indeed, as discussions have established, buried and daylighted rivers alike exemplify various facets of the Anthropocene. Their disrupted flows, material culverting and/or daylighted cityscapes now exist as geo-physical features that would linger on as traces of human intervention were our species to suddenly become decimated, disempowered or extinct. Indeed, it is illuminating to use the familiar tactic of imagining what archaeologists of the future (whatever species they might be) might make of buried rivers and their daylighted sections. If these future archaeologists could accurately date the historical points at which culverts had been constructed, and then sections subtracted and daylighting projects undertaken, they might perceive these as having been constructed, de- and re-constructed in a rapid temporal blur. They might also construe this as an enigmatic performance practice in which the transitions and temporal dynamics were an integral part of the phenomenon. Indeed, in an historical sense, there is much to support this contention. The construction and maintenance of urban landscapes involve structures and infrastructures being created, modified, obliterated by various factors (fire, war, deliberate demolition) and rebuilt in repetitive patterns. These arguably reached a peak of intensity – in the developed First World, at least – in what has been termed the 'great acceleration' of population growth, resource exploitation and pollution in the mid-late 20th century (McNeill and Engelke, 2014). While its first instances occurred during the peak period, daylighting (and blue-green rehabilitation, more generally) might be seen as deceleration phenomena and, thereby, to offer grounds for hope for areas of cities to be regenerated as cleaner, safer, more hospitable environments.

Developmental phases are different in the economically emergent countries of the Global South and, indeed, may not simply mirror sequences prominent elsewhere. In India for example, the picture regarding ongoing culverting, the maintenance of culverts and moves towards deculverting in some locations is more complex. Concerns about flooding of residential and industrial areas and of roads has resulted in considerable research on culverting technologies designed to ameliorate such events (e.g., Jadhav, Nimblakar, Paygude and Jadhav, 2024) and leading western culverting companies such as Inpipe Sweden have targeted India as a market for their products following the conclusion of an EU-India Free Trade Agreement in 2025 (Srivastava, 2025). At the same time, researchers such as Menon and Sharma (2022) have provided valuable perspectives for evaluating processes of daylighting and blue-green restoration with regard to waterways such as the Mullassery canal that flows through Kochi's CBD, the purposes of the channel and conflicting patterns of development along and over it. Unlike some of the more cosmetic daylighting schemes proposed for areas of European and North American cities, the researchers maintain a focus on urban functionality along with eco-aesthetics.

As various chapters have illustrated, there are complex agendas in play with regard to riverine heritage in urban environments. The sweeping away of previous rivers, riverscapes and tangible and intangible heritage elements associated with them is more easily accomplished than the restitution of these. One way of understanding this is to consider culverted flows such as The Fleet as repositories of eco-cultural (post)memories that have been serially modified by a succession of writers, creative artists, urban explorers and everyday urbanites. Indeed, it is the rivers' sealed-over forms that facilitate rich imaginative interpretations and imaginations of them. But daylighting does not necessarily release these elements into the ether and ethos of modern cityscapes (like the multiple ghosts released from containment boxes into Manhattan in the 1984 feature film *Ghostbusters*). Indeed, daylighting might be perceived to dispel or diminish such ghosts as revealed riverine flows cannot easily manifest the complex riverine heritages that might be associated with them.

Note

[1] See Hayward (2025) for discussion.

References

Akasegawa, G., Fujimori, T. and Minami, S. (eds) (1986) *Rojokansatsugakunyumon*. Chikuma Shobo, Tokyo, Japan.

Ankyo Maniacs, Takayama, H. and Yoshimura, N. (2020) Mizunonai mizube ni mieru-mono – Futari no ankyo mania no shisen. *Dobokugakkaishi* 105, 16–19.

Ankyo Maniacs, Takayama, H. and Yoshimura, N. (2022) Mizu wo kanji mizu wo aruku. Ankyo maniacs ga mitsumeru toshi no suimyaku. *Kenchikuzassi* 137, 12–13.

Ankyo Maniacs, Takayama, H. and Yoshimura, N. (n.d.) *Ankyo Maniacs*. Available at: https://www.ankyomaniacs.com/project (accessed 8 December 2025).

Argent St. George (2001) Principles for a human city. Available at: https://www.kingscross.co.uk/media/Principles_for_a_Human_City.pdf

Ashton, J. (1888) *The Fleet: Its river, prison and marriages*. T. Fisher Unwin.

Auckland Council (2008) Stream daylighting: Identifying opportunities for Central Auckland: Concept design. Available at: http://www.aucklandcity.govt.nz/council/documents/technicalpublications/tr2008-027%20stream%20daylighting%20identifying%20opportunities%20part%201.pdf

Barton, N. (1962) *The lost rivers of London: A study of their effects upon London and Londoners, and the effects of London and Londoners upon them*. Phoenix House – University of Leicester Press.

Barton, N. and Myers, S. (2016) *The Lost Rivers of London*. Historical Publications Ltd.

Billingsley, J. (2017) The city and country: Psychogeography as we see it. *Northern Earth* 150, 13–18.

Bishop, P. and Williams, L. (2019) *Planning, Politics and City-Making: A Case Study of King's Cross*. RIBA Publishing.

Boelens, R., Hoogesteger, J., Swyngedouw, E., Vos, J. and Wester, P. (2016) Hydrosocial territories: A political ecology perspective. *Water International* 41(1), 1–14.

Boelens, R., Hommes, L., Hoogesteger, J., Swyngedouw, E., Vos, J. *et al.* (2025) Hydrosocial territories: Imaginaries, materialities, and struggles over knowledge, order and meaning. *Water International* 50(5), 426–462. DOI: 10.1080/02508060.2025.2528261.

Bolton, T. (2011) *London's lost rivers*. MIT Press.

Bolton, T. (2019) *London's lost rivers Volume 2*. Strange Attractor Press.

Bonnett, A. (2009) The dilemmas of radical nostalgia in British psychogeography. *Theory, Culture and Society* 26(1), 45–70.

Brint, S. (2001). Gemeinschaft revisited: A critique and reconstruction of the community concept. *Sociological theory*, 19(1), 1-23.

Broinowski, B.A. (1945, April 21) I stroll the Tank Stream. *Sydney Morning Herald* 8.

Brutsaert, W. (2023) *Hydrology: An Introduction*, 2nd edition. Cambridge University Press.

Canberra Times (1981, April 18) Tank Stream Fountain features in new Herald Square 18.s.

Castonguay, S. and Evenden, M. (2012) Introduction. In: Castonguay, S. and Evenden, M. (eds) *Urban rivers: Remaking rivers, cities, and space in Europe and North America*. University of Pittsburgh Press.

Center for Climate Integrity (2025) Michigan climate impacts and costs. Available at: https://climateintegrity.org/uploads/media/CCI_Michigan_Impacts_Costs_2023.pdf

Chambers, F. (2015, January 10) Tracing the hidden River Sherbourne under Coventry. *BBC News Online*. Available at: https://www.bbc.com/news/uk-england-coventry-warwickshire-30446327

Charbonneau, R. (2000, Spring) Strawberry Creek II: Restoring the creek, 1987-1989 – a personal perspective. *Chronicle of the University of California* 19–26.

Charbonneau, R. and Resh, V.H. (1992) Strawberry Creek on the University of California, Berkeley campus: A case history of urban stream restoration. *Aquatic Conservation: Marine and Freshwater Ecosystems* 2(4), 293–307.

Chosunilbo (2025, October 10) Editorial: Cheonggyecheon innovation jolts ineffective politics, administration. *The Chosun Daily*, Available at: https://www.chosun.com/english/opinion-en/2025/10/10/ZKUWLMD3BBAXPFIKQIE7R5ZHCM/

City of Kalamazoo (2025) Arcadia Creek ("Phase O" Westnedge Avenue). Available at: https://www.kalamazoocity.org/Community/Projects/2025/Arcadia-Creek-Phase-0-Westnedge-Ave

City of Sydney (2025) Fountain, Australia Square Plaza between Pitt Street and George Streets, 1986. Available at: https://archives.cityofsydney.nsw.gov.au/nodes/view/673605

Cole, A., Healy, A. and Morel-Journel, C. (2022) Introduction to the role of urban narratives: Governing the (re)making of cities. In: Cole, A., Healy, A. and Morel-Journel, C. (eds) *Constructing Narratives for City Governance: Transnational Perspectives on Urban Narration*. Edward Elgar Publishing, pp. 1–16.

Connelly, A., Hanania, S. and Kiernicka-Allavena, J. (2019) Water, heritage and sustainability in practice: The cases of Rochdale and Wroclaw. *European Journal of Creative Practices in Cities and Landscapes* 2(1), 139–152.

Cooper, M., Chakraborty, A. and Chakraborty, S. (eds) (2018) *Rivers and Society: Landscapes, Governance and Livelihoods*. Routledge.

Cotoi, C., Iorga, A., Păduraru, R. and Georgescu, A. I. (2025). The lives and afterlives of the Dâmbovița: Between ghost and living river. *Environment and Planning E: Nature and Space*, 8(4), 1447–1464.

Cox, D. (2017, August 29) A river runs through it: The global movement to daylight urban waterways. *The Guardian*, Available at: https://www.theguardian.com/cities/2017/aug/29/river-runs-global-movement-daylight-urban-rivers

Damodaran, H. (2021) Spatial analysis exploring the influence of stream daylighting in building resilient urban form: A study of Zürich, Switzerland's 'Bachkonzept' for stream daylighting, University of Waterloo, Masters thesis.

Davies, H. (2013) *Following the Fleet – London's secret river*. Smashwords. Available at: https://www.smashwords.com/books/view/373464

Debord, G. (1958) Theory of the Dérive (K. Nabb, Trans.). *Internationale Situationniste* (2). Available at: https://www.smashwords.com/books/view/373464

Delibas, M. and Azime Tezer, A. (2017) 'Stream daylighting' as an approach for the renaturalization of riverine systems in urban areas: Istanbul-Ayamama Stream case. *Ecohydrology & Hydrobiology* 17(1), 18–32.

Douglas, C. (2025, September 17) Understanding place through history and geography: Walking the Fleet River. *Medium*. Available at: https://medium.com/@changebydesign/understanding-place-through-history-and-geography-walking-the-fleet-river-f1a944af24b8

Easterling, K. (2003) Subtraction. *Perspecta* 34, 80–90.

Eckenwiler, L.A. (2016). Defining ethical placemaking for place-based Interventions. *American journal of public health*, 106(11), 1944–1946.

Ellis, C. (2003) *The ethnographic I: A methodological novel about autoethnography*. Alta Mira.

Emmett, R.S. and Nye, D.E. (2017) *The environmental humanities: A critical introduction*. MIT Press.

Evans, G. (2018) Hydrocitizenship: concepts and insights from the lee valley, UK. In: Cooper, M., Chakraborty, A. and Chakraborty, S. (eds) *Rivers and Society: Landscapes, Governance and Livelihoods*. Routledge, pp. 202–219.

Falkenmark, M. (1977) Water and mankind: A complex system of mutual interaction. *Ambio* 6(1), 3–9.

Fathers, D. (2024) *London's Hidden Rivers: A walker's guide to the subterranean waterways of London*. Bloomsbury.

Fisiak, T., Hayward, P. and Fleury, C. (2024) Resurrecting riverine identity: Civic symbolism, ghostly traces and waterway restoration in 21st century Łódź (Poland). *Anthropocenes* 5(1), 1–16.

Fletcher, B.H. (2006) Arthur Phillip. In: *Australian Dictionary of Biography*, Vol. 2. Available at: https://adb.anu.edu.au/biography/phillip-arthur-2549

Flynn, H. and Brooks, A. (2020) The underground bridges of Rochdale. Available at: https://www.theskyliner.org/blog-1/river-roch-bridges

Foord, A.S. (1910) *Springs, streams and spas of London*. Frederick A. Stokes.

Fordahl, C. (2022) Community and the critique of technology: A revisionist account of an essential concept. *Capitalism Nature Socialism* 33(1), 71–89.

Fountain Workshop (n.d.) Pancras Square, King's Cross. Available at: https://www.fountains.co.uk/project/pancras-square

Gonzalez, C. (2026) Sediment as unintentional infrastructure: Decompositional politics and the disputed matter of Addis Ababa's urban riverbeds. *Environmental Humanities* 18(1), 187–204.

Gooden, R. (2019) Excursus three: Fleeting memories – bringing the Fleet back to life in St. Pancras. In: Adler, G. and Guerci, M. (eds) *Riverine Architecture and Rivers*. Routledge, pp. 135–142.

Groves, R. (1962, May 18) A Turbot in the Thames (review). *The Tribune* 10.

Halliday, S. (2001) *The Great Stink of London: Sir Joseph Bazalgette and the cleansing of the Victorian metropolis*. The History Press.

Harvard University Graduate School of Design (2025) Urban design case study archive: Cheonggyecheon Restoration Project. Available at: https://udcsa.gsd.harvard.edu/projects/9

Hasegawa, J. (2015a) Urban planning by obliteration of both waterways and opponents: The infilling of canals during the 1950s reconstruction of central Tokyo. *Urban History* 42, 483–506.

Hasegawa, J. (2015b) Tokyo's elevated expressway in the 1950s: Protest and politics. *Journal of Transport History* 36, 228–247.

Hasegawa, J. (2018) Debates on urban reconstruction through reclamation of traditional water scenery in 1940s Tokyo. *Planning Perspectives* 33, 29–52.
Hasegawa, J. (2023) The infilling and reclamation of inland waterways in Tokyo, 1945–1962. *Shima* 17(2), 93–120.
Hasegawa, J. (2025) Redeveloping Tokyo's Meiji Jingu Gaien area: The Metropolitan Government's city planning runs amok. *Asia-Pacific Journal Japan Focus* 23, 1–14.
Hasegawa, J. (2026) Transforming a Central Tokyo Park into a cutting-edge commercial complex: Miyashita Park, Shibuya. *Urban Planning* 11, 1–20.
Haycraft, C. (1962, March 16) Waterways (Book Reviews). *New Statesman* 381–383.
Hayward, P. (ed.) (2025) *Blue-Green rehabilitation: Urban planning, leisure and tourism in river cities*. CABI.
Hebbert, M. (2018) The long after-life of Christopher Wren's short-lived London plan of 1666. *Planning Perspectives* 35(2), 231–252.
Heise, U., Christensen, J. and Niemann, M. (eds) (2017) *The Routledge companion to the environmental humanities*. Taylor & Francis.
Herdt, A.J., Vanos, J. and Kim, J. (2017, January) Urban microclimate monitoring in Seoul, Korea: Fine scale summer heating along the Cheonggye Stream renewal project. In: *American Meteorological Society Meeting Abstracts*, p. 755 (Vol. 97).
Herrera-Franco, G., Montalván-Burbano, N., Carrión-Mero, P. and Bravo-Montero, L. (2021) Worldwide research on socio-hydrology: A bibliometric analysis. *Water* 13, 1283.
Hirsch, M. (2012) *The generation of postmemory: Writing and visual culture after the Holocaus*. Columbia University Press.
Honda, S. (2012) *Chikei wo Tanoshimu. Tokyo 'Ankyo' Sanpo*. Yosensha, Tokyo, Japan.
Honda, S. (2017). *Tokyo Ankyogaku*. Yosensha, Tokyo, Japan.
Honda, S. (ed.) (2021) *Ushinawareta Kawa wo Aruku. Tokyo 'Ankyo' Sanpo. Kaiteiban* [Revised edition]. Jitsugyononihonsha, Tokyo, Japan.
Honda, S. (2023) *Ushinawareta Kawa wo Yomu, Tsumugu, Tanoshimu. Tokyo Ankyogaku. Kaiteiban* [Revised edition]. Jitsugyononihonsha, Tokyo, Japan.
Honda, S. (2024) Ankyo ga tsunagu kako to ima. *Cemedine Style*. Available at: https://www.cemedine.co.jp/cemedine_reports/honda-ankyo.html (accessed 29 November 2025).
Honda, S. (n.d.) *Tokyo Ankyo Laboratory*. Available at: https://www.tokyoankyolabo.net/ (accessed 8 December 2025).
Honda, S., Takayama, H., Yoshimura, N. and Mitsuchi, T. (2017) *Hajimeteno Ankyo Sanpo. Mizunonai Mizube Wo Aruku*. Chikumashobo, Tokyo, Japan.
Hoobyar, P. (2002) *Daylighting and restoring streams in rural community city centers: Case studies*. National Park Service Rivers, Trails, and Conservation Assistance Program Seattle, Washington.
Hudson River Park (2020). Available at: https://hudsonriverpark.org/app/uploads/2020/07/HRPK_2020_OurWestSideStory.pdf
Hwang, K. (2005) Acknowledgement. In: Hwang, K., Byun, M., Lah, T.J. and Lee, S. (eds) *Cheonggyecheon Restoration Project: Conflict Management Strategies (KOTI Knowledge Sharing Report 22)*. The Korea Transport Institute, pp. 10–13.
Ingram, M. (2014) Washing urban water: Diplomacy in environmental art in the Bronx, New York City. *Gender, Place & Culture* 21(1), 105–122.
Inomata, H., Yokouchi, N. and Okada, T. (2000) 'Kotonaibukasenseibijigyo' nimiru unga no riyo nikansuru kenkyu. *Dobokukeikakugaku Kenkyukoenshu* 187–190. Available at: http://library.jsce.or.jp/jsce/open/00039/2000/23-1-0187.pdf

Irish, P. and Goward, T. (n.d.) Location: The Tank Stream. *Barabi: Sydney's Indigenous History*. Available at: https://www.sydneybarani.com.au/sites/the-tank-stream/

Ishihara, S. and Takasaki, T. (2015) Nihonbashigawa niokeru Shutokosokudoro no jokusenyo niitaru keii. *Tokyoto Dobokugijutusien Jinzaiikusei Senta Nenpo*. Available at: https://www.kensetsu.metro.tokyo.lg.jp/documents/d/kensetsu/000021817 (accessed 11 December 2025).

Ishihara, S., Takasaki, T., Kawamura, A. and Amaguchi, H. (2015) Tokyo Nihonbashigawa niokeru Shutokosokudoro no kokakozokettei niitaru haikei nitsuite. *Suimon Mizushigengakkai Kenkyuhappyokai Yoshishu* 28.

Ishihara, S., Kawamura, A., Takasaki, T. and Amaguchi, H. (2017) Shibuyagawa no kaishukeikaku no hensen to haikei nikansuru kenkyu. *Dobokugakkai Ronbunshu G (Kankyo)* 73, I_291–I_301. Available at: https://akira.fpark.tmu.ac.jp/paperlist/fullpaper/163.pdf (accessed 11 December 2025).

Ishihara, S., Kawamura, A., Takasaki, T. and Amaguchi, H. (2018) Nihonbashigawa niokeru Shutokosokudoro no judansenyo niitaru keikakukentokeii no kenkyu. *Dobokugakkai Ronbunshu G (Kankyo)* 74, 333–339.

Ishihara, S., Kawamura, A., Amaguchi, H. and Takasaki, T. (2019) Tokyo niokeru mizu to midori no nettowaku soshutsu to fukugaikakasen no saikaikyoka nikansuru ichikosatsu. *Suimon Mizushigengakkai 2019nendo Kenkyuhappyokai* 232–233.

Jadhav, N., Nimbalkar, P., Paygude, P. and Jadhav, S. (2024). Analysis and design of RCC Box Culvert. *SSRG International Journal of Civil Engineering*, 11(4), 54–64.

Jonson, B. (1692) On the famous voyage. *Epigrams* CXXXIV. Available at: https://hollowaypages.com/jonson1692epigrams.htm (accessed 20 September 2025).

Khirfan, L., Peck, M. and Mohtat, N. (2020) Digging for the truth: A combined method to analyze the literature on stream daylighting. *Sustainable Cities and Society* 59, 102225.

Kim, E. J. (2020). The historical landscape: Evoking the past in a landscape for the future in the Cheonggyecheon reconstruction in South Korea. *Humanities*, *9*(3), 113. https://www.mdpi.com/2076-0787/9/3/113

Kim, H. and Jung, Y. (2019) Is Cheonggyecheon sustainable? A systematic literature review of a stream restoration in Seoul, South Korea. *Sustainable Cities and Society* 45, 59–69.

King, J. (2025) About hidden hydrology. Available at: https://www.hiddenhydrology.org/about/

Kitson, J., Bucknum, M. and Meenar, M. (2023) Staging hydrophilic encounters – experiential methods for creating dialogic listening space. *Social & Cultural Geography* 24(5), 778–795. DOI: 10.1080/14649365.2021.1965196.

Lashua, B.D. and Baker, S. (2018) Urban subversion and mobile cinema: Leisure, architecture and the "Kino-Cine-Bomber". In: Dunlap, R., Harmon, J. and Rose, J.N. (eds) *The Politics of Leisure*. Routledge, pp. 49–62.

Lee, C.S., Lee, H., Kim, A.R., Pi, J.H., Bae, Y.J. *et al.* (2020) Ecological effects of daylighting and plant reintroduction to the Cheonggye Stream in Seoul, Korea. *Ecological Engineering* 152, 105879.

Lim, H., Kim, J., Potter, C. and Bae, W. (2013) Urban regeneration and gentrification: Land use impacts of the Cheonggye Stream restoration project on the Seoul's central business district. *Habitat International* 39, 192–200.

Lyons, S. (2017) Psychogeography: A way to delve into the soul of a city. The Conversation. Available at: https://theconversation.com/psychogeography-a-way-to-delve-into-the-soul-of-a-city-78032

Lyons, K. (2023) Rivers and Reconciliation: Elaborating the socioecological memory of war through science and arts-based practices. *Environmental Humanities* 15(1), 141–163.

Lythberg, B. and Hikuroa, D. (2020) How can we know Wai-Horotiu - A buried river? Cross-cultural ethics and civic art. *Environmental Ethics* 42(4).

Macphail, M. and Owen, T. (2018) What was growing along the Tank Stream Valley, Sydney Cove, in 1788? *Australasian Historical Archaeology* 36, 16–28.

Magata, H. (1996) Jidai wo utsusu toshi kasen. Tokyo no kubu kasen nitsuite. *Gesuibunkakenkyu* 8, 8–21. Available at: https://npo-jade.com/wp-content/uploads/2025/08/8-Magata.pdf (accessed 11 December 2025).

McEwen, L., Gorell Barnes, L., Phillips, K. and Biggs, I. (2020) Reweaving urban water-community relations: Creative, participatory river "daylighting" and local hydrocitizenship. *Transactions of the Institute of British Geography* 45, 779–801.

McIllwraith, J.F. (1952) The Tank Stream city of Sydney, N.S.W. *Australian Surveyor* 14(1–2), 37–54.

McNeill, J.R. and Engelke, P. (2014) *The Great Acceleration: An environmental history of the Anthropocene since 1945*. Bellknap Press/University of Harvard Press.

Menon, A.N. and Sharma, P. (2022) Reclaiming and rejuvenating urban water bodies: Case of Mullassery Canal, Kochi, Kerala. In: Gangwar, R., Agarwalla, A. and Sreekumar, S. (eds) *Infrastructure Development–Theory, Practice and Policy*. Routledge, pp. 52–59.

Mentz, S. (2009) Toward a blue cultural studies: The sea, maritime culture, and early modern English literature. *Literature Compass* 6(5), 997–1013. DOI: 10.1111/j.1741-4113.2009.00655.x.

Mentz, S. (2019) Swimming into the Blue Humanities. Available at: https://www.sydney.edu.au/sydney-environment-institute/events/past-events/2019/october/swimming-into-the-blue-humanities.html

Mentz, S. (2024) *An introduction to the blue humanities*. Routledge.

Metropolitan Expressway Company Limited (n.d.) Shutokosokudoro Nihonbashikukan chikakajigyo. Available at: https://www.shutoko.jp/ss/nihonbashi-tikaka/ (accessed 29 November 2025).

Ministry of Land, Infrastructure, Transport and Tourism (2018) Shutoko Nihonbashi chikakakentokai. Dai3kai (2018nen 7gatsu 18nichi) haifushiryo shiryo1. Gaisanjigyohi oyobi jigyosukimu nitsuite. Available at: https://www.mlit.go.jp/road/ir/ir-council/exp-ug/pdf03/02.pdf (accessed 29 November 2025).

Ministry of Land, Infrastructure, Transport and Tourism (n.d.) Kinnen no kou oyobi naisuihanran no jokyo, gesuidoseibi no genjo nitsuite. Available at: https://www.mlit.go.jp/mizukokudo/sewerage/content/001320996.pdf (accessed 11 December 2025).

Miyazawa, H. (2021) Kawatteikumono, kawaranaimono. Miyazawa Hio no taimuresu ni ainiiku. Dai6kai Polo Ralf Lauren no Big Oxford Shirt de ankyo sagashi. *Men's Non-No Web, 15 May*. Available at: https:www.mennono.jp.fashion.92423#shm-vp-1 (accessed 29 November 2025).

Mizbering (2026). *Mizbering*. https://mizbering.jp

Mkutu, K., Anderson, D.M., Lugusa, K., Owino, E.A. and Mkutu Agade, T. (2025) Scarcity, structures and scoundrels: Water-related conflicts in northern Kenya. *Water International* 50(7), 680–702.

Montagner, C. (2015, October 20) Projeto quer descobrir as centenas de rios que São Paulo enterrou. *Outra Cidade*. Available at: https://outracidade.com.br/projeto-quer-descobrir-as-centenas-de-rios-que-sao-paulo-enterrou/

Murphy, A. (1761) *An Ode to the Naiads of Fleet-Ditch*. Gale ECCO.

Nakamura, S. and Oki, T. (2009) 36toshin niokeru toshikasen haishi madeno keii to sono shiso. *Suikogakuronbunshu* 53, 565–570. Available at: http://library.jsce.or.jp/jsce/open/00028/2009/53-0095.pdf (accessed 29 November 2025).

Nakamura, S., Kimura, M., Noda, K. *et al.* (2020) Shakaisuimongaku—Sono Nihon deno tenkai ni mukete. *Suimon Mizushigen Gakkaishi* 33, 203–211.

National Cultural Flows Planning and Resource Committee (2017) Cultural flows: A guide for resource managers. Available at: https://www.mdba.gov.au/sites/default/files/publications/macquarie-castlereagh-wrp-cultural-flows-a-guide-for-water-managers.pdf

Nichols, W.J. (2014) *Blue mind: How water makes you happier, more connected and better at what you do*. Hachette UK.

Nihonbashi River Walk (2025) Oshirase. 11 June. Available at: https://www.nihonbashiriverwalk.jp/news/release_20250611/ (accessed 4 December 2025).

Oppermann, S. (2023) *Blue humanities: storied waterscapes in the Anthropocene*. Cambridge University Press.

Parker, C. (2016, January 17) Walking in water: The underground rivers of Tokyo. *The Tokyo Files*. Available at: https://thetokyofiles.com/2016/01/17/walking-on-water-the-underground-rivers-of-tokyo/

Pope, A. (1763) *The Dunciad Book II*. Available at: https://allpoetry.com/poem/8448455-The-Dunciad-Book-II.-by-Alexander-Pope

Poulos, N. (2021) *Essentials of Autoethnography*. American Psychological Association.

Quinn, R. (1936) The Tank Stream. (1936, December 9). *Australian Worker* 10.

Revkin, A. (2009, July 6) Peeling back pavement to expose watery havens. *New York Times*. Available at: https://www.nytimes.com/2009/07/17/world/asia/17daylight.html?pagewanted=1&ref=global-home

Rosbjerg, D. and Rodda, J. (2019) IAHS: A brief history of hydrology. *History of Geo- and Space Sciences* 10(1), 109–118.

Ross, A. and Chang, H. (2020) Socio-hydrology with hydrosocial theory: Two sides of the same coin. *Hydrological Sciences Journal* 65(9), 1443–1457. DOI: 10.1080/02626667.2020.1761023.

Rufus (1934, August 2) *The Tank Stream*. Sydney Morning Herald, p. 23.

Sass, O. and Viles, H. (2022) Heritage hydrology: A conceptual framework for understanding water fluxes and storage in built and rock-hewn heritage. *Heritage Science* 10, 66.

Sawayama, R. (n.d.) Rojo-Kansatsu-Gakkai. *Bijutsu Techo*. Available at: https://bijutsutecho.com/artwiki/46 (accessed 12 December 2025).

Scholte, T.I. (2020) Insite/outsite: The perpetuation of site-specific installation artworks in museums. Universiteit of Amsterdam doctoral thesis.

Seoul Solution (2014) Seoul Urban Regeneration: Cheonggyecheon Restoration and Downtown Revitalization. Available at: https://www.seoulsolution.kr/en/content/seoul-urban-regeneration-cheonggyecheon-restoration-and-downtown-revitalization

Seoul Solution (2021). Seoul Urban regeneration: Cheonggyecheon Restoration and Downtown Revitalization. Available at: https://www.seoulsolution.kr/en/content/seoul-urban-regeneration-cheonggyecheon-restoration-and-downtown-revitalization

Sheaf and Porters River Trust (2025). Available at: https://www.sheafportertrust.org

Shimomoto, E., Takayama, H., Yoshimura, N. and Nakamura, S. (2023) Kako hyakunen no Tokyo no kasen/suiromo no rekishikukandetabesu no kochiku. *Suimon Mizushigengakkai/Nihonsuimonkagakukai 2023nendo Kenkyuhappyokai Posutahappyo*. Available at: https://www.jstage.jst.go.jp/article/jshwr/36/0/36_186/_pdf/-char/en (accessed 29 November 2025).

Shoji, S. (1989) Shinsaifukkojigyo niokeru kasen unga keikaku. *Dai9kai Nihon Dobokushikenkyuhappyokaironbunshu* 9, 165–172. DOI: 10.11532/journalhs1981.9.165. (accessed 29 November 2025).

Shute, J. (2024) Cultural daylighting: Using stories of the past and present to reimagine Manchester's 'lost' River Irk. *Folklore* 135(4), 608–632.

Sibilla, S., Sciandra, M.C., Rosso, R. and Lamera, C. (2016) Hydraulic approach to Navigli canal daylighting in Milan, Italy. *Sustainable Cities and Society* 32, 247–262.

Sinclair, I. (2005) Museums of Melancholy. *London Review of Books* 27(16), 14–17.

Sinclair, I. (2013) *Swimming to Heaven: The lost rivers of London*. Swedenborg Society.

Sivapalan, M., Savenije, H.H. and Blöschl, G. (2012) Socio-hydrology: A new science of people and water. *Hydrological Processes* 26(8), 1270–1276.

Srivastava, P. (2025, 28 March) Inpipe Sweden AB set to expand into India amid EU-India Free Trade Agreement advancements. *Economic Diplomacy.in*. Available at: https://www.economicdiplomacy.in/post/inpipe-sweden-ab-set-to-expands-into-india-amid-eu-india-fta-advancement

Stevenson, C. Ian. (2024). Restoration's dark side: Gentrification and daylighting the Saw Mill River in Yonkers, New York. In Gordon, G. (Ed.) *Rewilding the urban frontier: River conservation in the Anthropocene* (pp 265–286). University of Nebraska Press.

Sustainability Directory (2026) *Anthropocene infrastructure Impact*. Available at: https://pollution.sustainability-directory.com/area/anthropocene-infrastructure-impact/

Sutton, J. (2020) Place and memory: History, cognition, phenomenology. In: Floyd-Wilson, M. and Sullivan, G. (eds) *Geographies of Embodiment in Early Modern England*. Oxford University Press, pp. 113–133.

Swift, J. (1710) A description of a city shower. Available at: https://www.poetryfoundation.org/poems/50578/a-description-of-a-city-shower

Sydney Water (2024a) Circular Quay Tank Stream work: Restoring the Tank Stream. Available at: https://www.sydneywater.com.au/water-the-environment/what-we-are-doing/projects-in-your-area/circular-quay-pump-station-and-tank-stream-work.html

Sydney Water (2024b) Review of Environmental factors: Tank Stream Reinstatement, Circular Quay. Available at: https://www.sydneywater.com.au/content/dam/sydneywater/documents/REF-Tank-Stream-repair-Circular-Quay.pdf

Sydney Water (2025) The Tank Stream. Available at: https://www.sydneywater.com.au/content/dam/sydneywater/documents/tank-streamheritage-fact-sheet.pdf

Takayama, H. (2016) Watashi ga ankyo ni iku riyu. In: Minagawa, N. (ed.) *Tokyo Suribachi-Chikei Nyumon Episode 14*. East Press, Tokyo, Japan. Available at: https://matogrosso.jp/serial/suribachi_r04/ (accessed 29 November 2025).

Takayama, H. (2022) Sairoku: B&B×suirojokansatsunyumonten. 'Kazoku/kokoro' no tana. Kokoro no naka no ankyo. *Mainichi ankatsu! Ankyo hanta*. 24 February. Available at: https://lotus62ankyo.blog.jp/archives/29933222.html (accessed 29 November 2025).

Takayama, H. (2025) Tsunagumono toshiteno *ankyou*. Ankyo maniacs, Ankyo ga tsunagumono. *Cemedine Style,* 20 January. Available at: https://www.cemedine.co.jp/cemedine_reports/es4jma00000097vl.html (accessed 29 November 2025).

Takayama, H. (n.d.) *Mainichi ankatsu! Ankyo hanta*. Available at: https://lotus62ankyo.blog.jp/ (accessed 8 December 2025).

Takayama, H. and Yoshimura, N. (eds) (2020) *Ankyo Paradaisu!*. Asahi Simbunsha, Tokyo, Japan.

Takayama, H. and Yoshimura, N. (2023) *Ankyo Ni Kakaru Hashi Kara Miru Machi.* Jitsugyononihonsha, Tokyo, Japan.

Takenouchi, S. (2015) 'Ankyo' ni hamari, 'ankyo' wo dendosuru. *Ad Studies* 53, 34–36. Available at: https://www.yhmf.jp/as/.assets/ads_53.pdf (accessed 29 November 2025).

Talling, P. (2011) *London's lost rivers*. Random House.

Tang, H.F., Lee, A.Y. and Hung, S.H. (2024) Does built environment and natural leisure settings with bodies of water improve human psychological and physiological health? *Landscape and Ecological Engineering* 20(4), 547–558.

Thornbury, W. (1878) The Fleet River and Fleet Ditch. In: *Old and New London Volume 2*. pp. 416–426.

Toso, T., Spooner-Lockey, K. and Hetherington, K. (2020) Walking with a ghost river: Unsettling place in the Anthropocene. *Anthropocenes* 1(1), 1–17.

Tsing, A.L., Bubandt, N., Gan, E. and Swanson, H. (eds) (2017) *Arts of living on a damaged planet: Ghosts and monsters of the Anthropocene*. University of Minnesota Press.

UCL River Fleet Restoration Team (2009). *Seeing the light: Rediscovering the historic River Fleet.* Available at: https://www.ucl.ac.uk/news/2009/jul/seeing-light-rediscovering-historic-river-fleet

United Kingdom Parliament (1858) The Metropolis Management Amendment Act. Available at: https://www.legislation.gov.uk/ukpga/Vict/21-22/104/contents/enacted

Urbex (2025). Available at: https://urbexhub.com/the-history-of-urbex/

Vallerani, F. and Visentin, F. (eds) (2018) *Waterways and the cultural landscape*. Routledge, London, New York.

Visentin, F. (2024) *Geografie d'acqua: Paesaggi ibridi*. Marsilio.

Wang, Y., Borthwick, A. G. and Ni, J. (2022). Human affinity for rivers. *River,* 1(1), 4-14.

Wantzen, K. (2022) River culture: How socio-ecological linkages to the rhythm of the waters develop, how they are lost, and how they can be regained. *The Geographical Journal* 190, 1–16.

Wantzen, K.M., Piednoir, T., Cao, Y., Vazhayil, A.M., Tan, C. *et al.* (2022) Back to the surface–Daylighting urban streams in a Global North–South comparison. *Frontiers in Ecology and Evolution* 10, 838794.

Watts, P. (2005) Inside the Fleet: Exploring London's lost rivers. *Time Out*. Available at: https://greatwen.com/2011/01/27/inside-the-fleet-exploring-londons-lost-rivers/

Wesselink, A., Kooy, M. and Warner, J. (2017) Socio-hydrological and hydrosocial analysis: Toward dialogues across disciplines. *Wiley Interdisciplinary Reviews: Water*, 4(2), e1196.

Westling, E. L., Surridge, B. W., Sharp, L. and Lerner, D. N. (2014). Making sense of landscape change: Long-term perceptions among local residents following river restoration. Journal of Hydrology, 519, 2613-2623.

Whitehouse, A.C. (2024) *Korean nationalism and stream ecology in the Cheonggyecheon Museum*, Master of Arts thesis. University of Utah.

Wild, T.C., Dempsey, N. and Broadhead, A.T. (2019) Volunteered information on nature-based solutions — dredging for data on deculverting. *Urban Forestry & Urban Greening* 40, 254–263.

Yang-heok, K. (2025) Seoul enhances flood preparedness with five new rainwater basins and radar systems. *ChosunBiz*. Available at: https://biz.chosun.com/en/en-society/2025/05/12/7DGSSX4QVFGNRIGXNXM3E63KU4/

Yoon, Y. (2022) Cheonggyecheon Restoration Project: The politics and implications of globalization and gentrification. *Dartmouth Undergraduate Journal of Politics, Economics and World Affairs* 1(1), 8. Available at: https://digitalcommons.dartmouth.edu/cgi/viewcontent.cgi?article=1009&context=dujpew

Yoshimura, N. (2015) Mizunonai mizubekara... *'Ankyo'* no tanoshimikata. Dai6kai Hamachogawa to Ryukangawa. *Mizbering* 13 March. Available at: https://mizbering.jp/archives/13417 (accessed 6 December 2025).

Yoshimura, N. (2020a) Bunjin totomoni yoru no moso *ankyo* sanpo: Kanigawa to Rokkenbori, Gokenbori. In: Takayama, H. and Yoshimura, N. (eds) *'Ankyo Paradaisu!*. Asahi Sinbunshuppan', pp. 182–189.

Yoshimura, N. (2020b) Jirei ni miru, toshikaihatsu to ankyo no tatakai no rekishi. Tamagawajosui to Miharabashi. In: Takayama, H. and Yoshimura, N. (eds) *'Ankyo Paradaisu!*. Asahi Sinbunshuppan', pp. 76–79.

Yoshimura, N. (2025) Tsunagumono toshiteno 'ankyo'. Ankyo maniacs, Ankyo ga tsunagu mono. *Cemedine Style*. Available at: https://www.cemedine.co.jp/cemedine_reports/es4jma00000097vl.html (accessed 29 November 2025).

Yoshimura, N. (n.d.) *Ankyo sanpo*. Available at: http://kaeru.moe-nifty.com/ankyo/ (accessed 8 December 2025).

Yoshimura, N. and Takayama, H. (2015) *Ankyo Maniakku!* Kashiwashobo, Tokyo, Japan.

Yoshimura, N. and Takayama, H. (2021) *Machiaruki ga Tanoshikunaru Suirojokansatsunyumon*. Kadokawa, Tokyo, Japan.

Yoshimura, N. and Takayama, H. (2024) *Ankyo Maniakku! Zohoban*. Chikumashobo, Tokyo, Japan.

Zambiasi, T. (2025) Financialization and water crisis in São Paulo (2014–2015). *Geoforum* 167, 104464.

www.ingramcontent.com/pod-product-compliance
Lightning Source LLC
LaVergne TN
LVHW050437260926
848956LV00008B/30